The Beginnings of the Nobel Institution

The Beginnings of the Nobel Institution
THE SCIENCE PRIZES, 1901–1915

Elisabeth Crawford
Centre National de la Recherche Scientifique
Groupe d'Etudes et de Recherches sur la Science
Paris

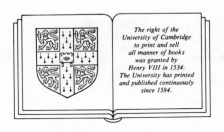

The right of the
University of Cambridge
to print and sell
all manner of books
was granted by
Henry VIII in 1534.
The University has printed
and published continuously
since 1584.

CAMBRIDGE UNIVERSITY PRESS

CAMBRIDGE
LONDON   NEW YORK   NEW ROCHELLE
MELBOURNE   SYDNEY

& EDITIONS DE LA MAISON DES SCIENCES DE L'HOMME

PARIS

Published by the Press Syndicate of the University of Cambridge
The Pitt Building, Trumpington Street, Cambridge CB2 1RP
32 East 57th Street, New York, NY 10022, USA
296 Beaconsfield Parade, Middle Park, Melbourne, 3206, Australia
and
Editions de la Maison des Sciences de l'Homme
54 Boulevard Raspail, 75270 Paris Cedex 06

First published in 1984

Printed in the United States of America

Library of Congress Cataloging in Publication Data
Crawford, Elisabeth T.
The beginnings of the Nobel institution.
Bibliography: p.
    Includes index.
1. Physics – Awards – History.   2. Chemistry – Awards –
History.   3. Nobel prizes – History.   4. Nobelstiftelsen –
History.   I. Title.
QC28.C73   1984   507'.9   84-5844
ISBN 0 521 26584 3
ISBN 2 7351 0086 3 (France only)

Contents

vi   *Contents*

Acknowledgments

I want to thank, first of all, the Royal Swedish Academy of Sciences for making this study possible by giving me access to archival materials concerning the Nobel prizes in physics and chemistry from 1901 to 1915. The Academy committee (Nobelarkivkommittén or NOAK for short) that oversees the archives has helped me in every possible way. I am particularly grateful to two members of the committee, Dr. Wilhelm Odelberg and Professor Bengt Nagel, who have given me both encouragement and friendly counsel.

I am equally grateful to Dr. Stig Ramel, executive director of the Nobel Foundation, who helped me gain access to the archives and obtain the small grant from the Foundation that permitted Professor Roy MacLeod and me to carry out the first inventory of materials in the archives. The Nobel Foundation has kindly given its permission to reprint certain sections from the article I wrote with Robert M. Friedman, "The prizes in the context of Swedish science" in C. G. Bernhard, E. Crawford, and P. Sörbom (eds.), *Science, technology, and society in the time of Alfred Nobel* (Oxford, Pergamon Press, 1982), © 1982 The Nobel Foundation.

I have been fortunate, during the five years it has taken me to complete this study, to be able to devote almost all my time to research and writing. I owe this luxury to my employer, the Centre National de la Recherche Scientifique (CNRS), which also has provided me with research assistance. I am grateful to my research director at the CNRS, M. Gérard Lemaine, who has encouraged and supported me throughout this project. The project was carried out as part of the research program of the Groupe d'Etudes et de Recherches sur la Science (affiliated with the Ecole des Hautes Etudes en Sciences Sociales and the CNRS), headed by Gérard Lemaine.

Several organizations have provided material or other support during the different phases of this study. I owe a large debt of gratitude to the Stiftung Volkswagenwerk, which put up the money for the microfilming of the bulk of materials in the Nobel Archives. The Office for the History of Science and Technology at the University of California, Berkeley, has taken on the computer analysis of nominators and nominees for the Nobel prizes in physics and chemistry, the first results of which are presented in this book. I am grateful to Professor John Heilbron, director of the Office, and Ms. Jacqueline Craig for collaborating with me on this part of the study. I am also indebted to the organizations that have provided me with travel grants, specifically the CNRS, Stiftung Volkswagenwerk, the Maison des Sciences de l'Homme, and the Association Franco–Suédoise pour la Recherche.

To gain insight into the early history of the Nobel institution, I have had to consult many sources other than the official records. The correspondence of a large number of Swedish scientists have been particularly important for my work. Most of these correspondences I found at the Library of the Royal Swedish Academy of Sciences, where I always received a warm welcome and precious help. I am grateful to the staff at several other institutions for access to materials in their possession and for their assistance: in particular the Mittag-Leffler Institute, the Uppsala University Library, the Niels Bohr Library, and the Center for History of Physics of the American Institute of Physics. I am also indebted to Professor Albin Lagerqvist and Docent Anders Bäcklin for giving me access to the minutes of the Physical Societies in Stockholm and Uppsala. Permission to reprint portions of the papers of T. W. Richards has kindly been granted me by the Harvard University Archives. The same courtesy was extended to me by the Carnegie Institution of Washington, D.C., with respect to the papers of G. E. Hale.

Many colleagues and friends have contributed to my work, each in their own way; unfortunately, it is not possible to thank them all individually. I deeply regret that the two people who did the most to get me started on this project – Professors Stein Rokkan and Sten Lindroth – are no longer with us. Among those to whom I want to give special thanks are the following: Robert Marc Friedman for many stimulating conversations, much help in locating archival materials, and useful suggestions with respect to the drafts of this manuscript; Terry Shinn and John Heilbron for having read and criticized the final draft; Mary Jo Nye for letting me partake of her deep knowledge of atomistics; Anna-Lisa Arrhenius-Wold for helping me locate documents in the Arrhenius Collection and for sharing her memories of her father; and Alan Duff for having translated so elegantly the long quotation from Henri Poincaré's letter of nomination in the section "The campaigns of Mittag-Leffler" in Chapter 5. (All the other translations from Swedish, French, or German into English are my own.)

Special thanks are due to Robin Sutcliffe and Mary Barnett who typed the different drafts and to Anne Gjesdal who removed many errors from the bibliography.

Finally, I want to register my debt, but I would not know how to begin to repay it, to my parents who have cared for me during my frequent visits to Stockholm and to my husband and son who have borne my many absences cheerfully and never questioned the worthwhileness of my work.

Introduction

During the more than eighty years of their existence, the Nobel prizes have become an institution. Early on, the elaborate structures that were set up for making the awards and administering the fund gave the prizes the permanence that is one of the defining characteristics of an institution. Over the years, the prizes have also taken on important social functions (this being another characteristic of an institution), particularly in the sciences, where they have become the main symbol of the system whereby scientists are offered recognition and reward for excellence by their peers. This, of course, results from the position that the prizes occupy at the very top of the hierarchy of honorific awards that has grown up with modern science.[1]

With the benefit of hindsight, it is easy to regard the present important position of the Nobel institution, particularly in the sciences, as inevitable. Yet the first and most significant step taken to set up the institution – the will of Alfred Nobel – did not in itself presage an institution, let alone an important one. (For the text of the relevant portion of the will, see Appendix A.) What Nobel instituted in his will (1895) were five prizes – in physics, chemistry, medicine or physiology, literature, and peace – to be awarded by four institutions: two academies (the Royal Swedish Academy of Sciences and the Swedish Academy [of literature]), one teaching institution (the Caroline Medico–Chirurgical Institute, known as the Karolinska Institute), and one legislative body (the Norwegian Storting or parliament).

To create the Nobel institution it was necessary for these four organizations to be bound by common rules and procedures so as to distinguish their role in awarding the prizes from the ones they normally performed and to prevent this role from being subsumed under these other functions. The statutes of the Nobel Foundation (promulgated in 1900) accomplished this. In addition, they provided for the new organs that were set up to implement Nobel's wishes – in particular, the Nobel Foundation as the corporation managing the prize fund. In this manner, the institution came to represent much more than its original constituent parts – the prizes and the prize awarders – since the statutes provided the institution with an overall structure and also guaranteed its permanence.

On one level, the Nobel institution can be seen as encompassing all the prizes: This global view was the basis for the significance that the public attached to the prizes from the beginning. On another level, because of the important role that the prizes in science and medicine played (and have continued to play) in the creation of the institution, it can be viewed as a *scientific institution*. The fact that the Royal Academy of Sciences was entrusted with *two* prizes, in the neighboring disciplines of physics and chemistry,[2] made it – and its prizes – a particularly important part of the early Nobel institution. It is on this part that I have chosen to focus the present study. The following close examination of how these prizes were awarded (within the framework of the general rules set for the functioning of the institution), and how they fitted into disciplinary developments as well as the scientific culture of the time, tries first of all to reconstruct the initial stage in the development of the institution. It also shows, I hope, how this stage contained many of the sources of its subsequent growth into a significant scientific institution.

Needless to say, such a reconstruction would not have been possible without access to the Nobel Archives – in the present case those of the Royal Swedish Academy of Sciences and its Nobel Committees for Physics and Chemistry. As a result of the new rule, instituted in 1974, making archival materials dating back fifty years or more available for historical research, I was given the opportunity to study the inner workings of the prize selections at the Academy* during the formative years. I have chosen to focus rather strictly on how this part of the institution developed in its

---

* Throughout the book, references in the text to "the Academy" and "the Academy of Sciences" are to be understood to mean the Royal Swedish Academy of Sciences. To avoid confusion between the Academy and other academies discussed, particularly in Chapter 1, the French Academy of Sciences will be referred to as "the Paris Academy" or "the academy" when its full name is not used. Likewise the Academy of Sciences of Berlin will figure as the "Berlin Academy."

first fifteen years – that is, before the First World War brought about a hiatus in the annual awarding of the prizes. This means that I have refrained from the free-ranging discourse on the role of the science prizes in general and over a longer period that has been so important in the literature on the Nobel prizes.[3] I feel that a narrower approach is justified, first, because the first fifteen years is too short a period to assess the significance of an institution, and, second, because there will be other studies based on materials in the Nobel Archives.[4] Once these have started to accumulate, I hope that such assessments will be placed on a more solid foundation of historical evidence than has been the case up to now.

The dictum "What does happen is not necessarily what had to happen" is particularly pertinent to the history of the Nobel institution in the sciences because of the fame and prestige the prizes have acquired over the years of their existence. At present, the Nobel prize occupies a unique position in the reward system of science in that it is as well known among the general public as it is to the scientific community. It also carries the distinction of being the only award of its kind that is regularly used to indicate the importance of a scientist or a discovery, not only those honored by the prizes but also the select group of scientists and works that are considered as being "of Nobel class" – *les nobélisables*, as they are known in French. That this is so is largely due to the way successive generations of laureates have contributed their prestige to the institution.[5] Over the years, their achievements have endowed the prize with an incremental value in a manner somewhat analogous to the contribution that each crowned head made to the institution of the monarchy in a bygone era.

Many of the aspects that have made the Nobel institution an important scientific institution were present from the beginning, but clearly they could not be the ones outlined above. What *did* happen during the early years was that Nobel's vague wishes were given substance and form, initially in the statutes and later by the statutory rules being put into practice in the selection of prize-winners. This was the work of the small group of Swedish scientists who formed the core of the institution and for whom Nobel's will represented an unprecedented opportunity to give an international dimension to Swedish science. Not surprisingly, Svante Arrhenius, who was the most internationally minded of Swedish physicists and chemists active at the time, was the chief actor in this group. For Arrhenius and his friends, the awarding of the prizes was part of a larger vision: that of an institution that would embrace several new research institutes (the Nobel Institutes) and also benefit existing institutions (principally the science faculties in Stockholm and Uppsala) by bringing them into contact with developments outside Sweden. It was this practical view that gave the institution a *raison d'être* over and above the awarding of the prizes and also made it, first and foremost, a Swedish scientific institution.

It is clear that, in the early years, the prize decisions depended heavily on the internal dynamics of the group of Swedish scientists who served on the Nobel Committees for Physics and Chemistry, since there was neither a tradition to guide them in making such decisions nor strong links with scientific centers abroad. It seems reasonable to assume that their judgment both of what specialties should receive consideration within the general fields designated for the awards, and of what specific works should be rewarded, were influenced by what they themselves considered important in these fields. Opinions are shaped by knowledge. Hence, not unexpectedly, the works considered for the prizes were those familiar to committee members because they formed part of their own areas of scientific interest. Since the latter were often institutionally based, the prize decisions were influenced by intellectual and institutional rivalries peculiar to Swedish science.

For the Nobel institution to develop into an important scientific institution required, however, that it acquire a constituency of interested scientists outside Sweden and that it find a place in the international disciplines of physics and chemistry. In the early years, this process paralleled that described in social histories of other scientific institutions,[6] since it was largely by extending outward from the core of Swedish scientists and by creating a participatory interest on the part of other groups that the institution eventually came to have a hold on important segments of the international scientific community. Even before the prize decisions had started to make an impact, the Nobel institution in this broader sense had come into existence through the statutory provision that candidates for the prizes be officially designated by specially invited nominators. Among these two "ready-made" and partially overlapping constituencies, the interest of the candidates was, of course, not in doubt because of the gains that they and their institutions stood to make if they were selected. But that of the nominators, too, was maintained, normally by the personal and professional ties that linked them to the nominees. These links had the greatest impact on the prize decisions when they involved Swedish scientists, and here the widest-ranging influence was that exerted by Arrhenius through his international network. That academic research in physics and chemistry (which produced most of the work proposed for the prizes) was practiced on an extremely small scale compared with the international scientific enterprise of today was a major factor making for participatory interest. Of the approximately one thousand physicists active in Europe and North America early in the century,[7] between one-fourth and one-third probably figured either as candidates or as nominators for the physics prize.

But apart from social and institutional considerations, the prizes also had to find their place in the *science* of physics and chemistry and in the scientific culture of the time. With respect to the latter, they broke new ground by being the first truly international prizes. At the same time, however (and this was more important for their recognition), they could be ranged with the innumerable honors that scientific societies in many countries were in the habit of bestowing on their own nationals as well as on eminent foreign scientists. However, it is with respect to the kinds of science and scientists that were honored by the prizes that the change in perspective on the institution that can be gained through studying its *internal* history during the early years becomes particularly important. This is so because the tendency to use the Nobel prize as the touchstone of excellence in science, referred to earlier, not only relates to the present but also has a strong hold upon the history of science. Invoking the Nobel prize in historical accounts is a form of shorthand: Simply mentioning that a discovery was recognized by the award of the prize is often enough to indicate its reception by the scientific community. Furthermore, invoking the Nobel prize in such accounts gives an impression of the inevitability of success befalling individual scientists: preselected "great men." Needless to say, this process of codification has been facilitated by the secrecy that until recently has surrounded the selection of prizewinners.

By contrast, when one considers the candidates and not just the laureates, one finds that there were many alternatives to the choices made and a fair amount of muddling over these. That the field could nevertheless be narrowed down to one final choice that generally met with the approval of other scientists was most often due to the fact that the choices were made in the context of prevalent specialties in physics and chemistry and, furthermore, reflected the judgments that specialty groups had already passed on the works or scientists proposed for the prizes.

In general, the drama of science in the making had been played out long before the works reached the prize juries. For this reason, only faint echoes were heard of the ongoing intellectual struggles of theoretical physicists who were giving new meaning to basic physical concepts in the first dècade of the century. Rather, insofar as the physics prize highlighted major discoveries (those of X-rays, radioactivity, and the electron, to mention only a few), these were primarily in *experimental* physics, which was not only the prevalent orientation at the turn of the century, but also the one which, in the opinion of contemporary observers, had brought the most significant advances. The context of awarding the prizes was not that of discovery per se, then, but of justification, not only of discoveries but of their importance for given specialties and fields. Formally, this was linked to the statutory provision that only published works and hence *public knowledge* could be considered for the prizes (making the secrecy surrounding the process of prize selection somewhat incongruous). Informally, this resulted from the way members of the institution (in the narrow and wider sense) took their cues from the prior acceptance of works proposed for the prizes.

OUTLINE OF THE BOOK

There are three major subdivisions in this book. The first (Chapters 1–3) treats elements of what might be termed the prehistory of the institution. It gives crucial background on the Nobel prizes in relation to preexisting reward systems, on the Swedish scientific community around 1900, and on the drafting of the statutes of the Nobel Foundation. The second part (Chapters 4–6) concerns the internal history or private side of the institution; it analyzes how prizewinners in physics and chemistry were selected during the period 1901–1915. The third part (Chapter 7) discusses the Nobel prize institution in its interactions with the public and the scientific community, or what might be called the public face of the institution in the early years.

In Chapter 1 I examine the national reward systems that were precursors of the Nobel prizes – in particular, the tradition to reward and encourage scientists by offering monetary awards that had evolved in the French Academy of Sciences. In this respect, the Nobel prizes continued a tradition, but as the first truly international prizes of modern times they represented a significant break with the past. Yet more important for the special status that they came to acquire in the eyes of the public were the large sums of money involved. With each prize representing 150,000 crowns[8] (when first awarded in 1901), they surpassed all awards of this kind created previously.

These and other unique qualities of the Nobel prizes did not emerge directly from Alfred Nobel's will. Indeed, his testament proved embarrassingly vague on so many points that putting it into operation was a long process, lasting from 1896 to 1900. In this initial phase, Swedish scientists shaped the institution by actively participating in the negotiations over the statutes. In order to learn more about the institutions that Nobel had named to award the prizes in science and medicine, and about the scientists (and their institutions) who would constitute the prize juries in physics and chemistry, an introduction to the Swedish scientific community in international perspective is provided in Chapter 2. Chapter 3 describes the events that led to the promulgation of the statutes of the Nobel Foundation in 1900.

In Chapter 4 I begin to examine decision making about the prizes in physics and chemistry. In this chapter, the relative weight of the system for proposing candidates for the prizes in the overall process of decision making is assessed through tabulations of the nominations that candidates and Nobel prizewinners received from different categories of nominators. These tabulations show that the convergence of nominators' opinions on the candidates who became prizewinners dissolved about midway through the period studied here. For the remainder of the period, only a few candidates (M. Planck, H. Poincaré, and W. Nernst in particular) received strong support from the nominators, but for a number of reasons their awards caused problems both in the committees and in the Academy of Sciences. The necessity of choosing from among several minority candidates increased the importance of the role of individual members of the Nobel Committees for Physics and Chemistry as well as of the evaluations carried out by them. In Chapter 5 I examine the decisive importance of Arrhenius in shaping the decisions and hence the international standing of the prizes. To a considerable extent his influence, as well as that of Gösta Mittag-Leffler, Sweden's best-known mathematician, depended on their activity in international networks and on the interest in the prizes that they created among members of such networks. By considering the actions they took to promote or block given candidacies in relation to these networks, a qualitative dimension is added to the primarily quantitative analysis of the nominating system of Chapter 4.

Notwithstanding the power of personalities, their role was influenced and in many ways tempered by the rules and procedures that governed the decisions concerning the prize awards. Chapter 6 examines how the formal rules laid down in the statutes were interpreted and supplemented by ad hoc decisions. In this manner, procedures were established that placed the Nobel committees in control at the same time as they were made accountable to the Academy of Sciences through their obligation to justify their recommendations and put on the table the evaluations leading up to these. This was clearly important for the legitimation of the decisions, as was the ambition to have the awards meet the criteria of the will and the statutes. To elucidate this latter point, I give special attention to the interpretations of the wording stipulating that the prizes be given for *discoveries and improvements* made *during the preceding year* and conferring *the greatest benefit on mankind*.

Statutory rules and criteria, however, were not sufficient to allow the committees to make a final choice, especially in the absence of guidance from the nominating system. In the second part of Chapter 6 I examine how the committees arrived at the consensual mode of decision making that was the major way of ensuring that the prize selection process would operate smoothly. This demanded not only that the committees overcome the difficulties posed by inherent structural features of Nobel prize decisions (in particular, the fact that among the many candidates proposed for the prizes only one or two could be chosen as the final prizewinners), but also that they resolve conflicting views among the members and/or the nominators as to what constituted discoveries of significance for physics and chemistry.

In Chapter 7 I discuss the recognition given to the institution in its early years, in both the popular and scientific press. There is no doubt that the public took an intense interest in all the prizes from the beginning, but particularly in those for literature and for peace. Although the science prizes also gained public standing, primarily through the rewarding of Pierre and Marie Curie for their discovery of radium, the process whereby these prizes would become prestigious awards in *the reward system of science* was only beginning during the period studied here. Nevertheless, even at this early stage some of the sources of their subsequent fame and prestige are apparent. I have tried to sketch these out, especially when they concern the use of the prizes to legitimate new fields. An Epilogue describes the winding down of the process of prize selection as a result of the First World War and offers an appraisal of what had been accomplished in the first fifteen years.

Finally, some definitions may help guide the reader through the maze of prizes, medals, and grants, the history of which will be sketched out in the first chapter. In a general sense this study concerns the *reward system of science*, this being the comprehensive term used in the sociology of science to designate all the ways and means by which scientists are rewarded for their work, usually by their peers. Prizes, medals, and grants – the three recurrent terms of the first chapter – are specific rewards in science and hence form a subset of the overall reward system. As a general rule, *prizes* can be defined as *rewards carrying monetary awards*. In addition to their monetary value, prizes also have a symbolic one, since they also confer honor and prestige on their recipients. *Medals* are *rewards that generally do not carry cash awards* and hence have no monetary value (unless they are melted down and/or sold). In contrast to the prizes and medals, which are given out for work already accomplished, *grants* are *awarded to enable scientists to carry out projected work*. It should be added that this last definition is the one that currently applies to the term *grant*. As will be shown in the subsequent description of the historical development of prizes, medals, and grants, when the Nobel prizes were instituted in the late nineteenth century, prizes and grants were more similar in both purpose and use than they are at present.

1

Precursors to the Nobel prizes in the sciences

To place the Nobel prizes in their proper historical setting it is necessary to retrace briefly the growth and changing functions of prizes and medals in the eighteenth and nineteenth centuries. The use and significance of prizes and medals fall into two distinct periods. In the first, extending from the early eighteenth to the middle of the nineteenth century, prizes were important chiefly as a means to reward successful entrants in the competitions organized by academies on the European continent, primarily those of Paris and Berlin. In the second period (1850–1915), the link with competitions was broken, and prizes, as was already the case with the medals of the Royal Society of London, were used to reward scientific achievement in general or work to which the donor of the prize or medal attached particular importance. During this period, the number of prizes and their monetary value both grew significantly; they also came to be regarded, by donors as well as by the institutions awarding them, as a means to stimulate and guide future work. This development was most pronounced in the French Academy of Sciences, where an elaborate system of prizes and grants emerged during the latter half of the nineteenth century. This system may well have influenced Nobel, who made Paris his home as well as the base of his wide-ranging business activities from about 1870 until his death in 1896.

THE PRIZE COMPETITIONS OF THE SEVENTEENTH AND EARLY EIGHTEENTH CENTURIES

The idea of rewarding scientists for their work is bound up with the emergence of science as a distinctive social activity practiced by those whose special abilities and training made them adept at unraveling the mysteries of nature. At the outset, this idea probably did not amount to more than providing a livelihood for those having such abilities and ambitions – for example, the stipends or payments in kind that "scientists" (long before this term had gained currency) received from monarchs or feudal princes in Renaissance times. They also indicate that even at this time, science was perceived as a means to enhance the power and prestige of the ruler. Yet while science then had both its patrons and rewards, it was not until scientists set up their own societies, aided by royal patronage, that such rewards took on new forms and became differentiated with respect to both the types of reward offered and the categories of scientists or scientific work that benefitted from these. The emergence of the awards that are the chief precursors of the Nobel prizes followed the foundation of two major scientific societies in the seventeenth century: the Royal Society of London (1660) and the French Academy of Sciences (1666). In the first two decades of the eighteenth century, these societies received gifts that set the reward system on the course it was to take in each society for the next century and a half.

In 1709, the Royal Society received a legacy of 100 pounds from Sir Godfrey Copley, one of its members, to be laid out on experiments or otherwise used for the benefit of the society. It decided, in 1736, to use the gift to institute a medal "to be bestowed on the person whose experiments should be best approved," thereby creating "a laudable emulation ... among men of genius ... who, in all probability may never be moved for the sake of lucre."[1] By contrast, the bequest made to the French Academy of Sciences in 1715 by Count Jean Rouillé de Meslay introduced the idea of encouraging scientists to find solutions to specific problems by offering monetary rewards. The two areas that preoccupied the donor and where he felt that learned men would be able to carry forward his "humble reflections" were, first, the movements of planets and other celestial bodies and, second, the accurate determination of longitudinal position at sea. In accordance with the donor's wishes the academy decided, in procedural rules adopted in 1719, to hold annual prize competitions in these areas. While the formulation of questions and the judging of essays represented new work for the academy, it was not without its rewards since the donor had had the foresight to set aside a certain sum to cover both the cost of announcements and the paying out of stipends or *épices* to the judges.[2]

The example set by Count Rouillé de Meslay was eventually followed by other donors and scientific societies. In 1744, the Academy of Sciences of Berlin instituted an annual prize competition that would soon attract as much or more attention among European philosophers of nature as that of the Paris Academy. The role of these competitions in stimulating and orienting scientific and philosophical work in the reign of Frederick the Great (1740–1786) has been eloquently described by Harnack in his history of the Academy of Sciences of Berlin:

> In a time when the energies and the organization for large scientific undertakings – with the exception of those in astronomy – were still lacking, the prize competitions announced annually by the academies in Europe became objects for scientific rivalries and the criterion for the standing and acumen of scientific societies.... This was so because specialities were most often disregarded and the themes chosen for competitions were either those that required perfect insight into the state of an entire discipline and its furtherance with respect to critical points, or those that posed a fundamental problem. The prize competitions constituted the lever by which the different sciences were raised one step higher from one year to another; in addition, they were important for universalizing and unifying science. The questions were addressed to learned men all over Europe and were communicated throughout the scientific world. The suspense surrounding the announcement of the question was, in fact, larger than that of the answer, for it was in the formulation of the question that mastery was revealed. The invitation was not addressed to young recruits of science but to the leaders who eagerly answered the call to contest. The foremost thinkers and learned men – Euler, Lagrange, d'Alembert, Condorcet, Kant, Rousseau and Herder – all entered the arena. This circumstance which may seem quite strange today requires special explanation. This latter ... resides in the fact that the learned man of the 18th century was still a *Universalphilosoph.* His mind could discern an abundance of problems in different scientific areas which all seemed equally attractive and enticing. Which one should he attack? At that moment, the Academy came to the rescue with its prize competitions. It presented him with a given theme and assured him a universally interested audience.[3]

The number of treatises submitted in prize competitions during this period testifies to the interest these evoked: In one year (1780), the theme set by the Berlin Academy attracted forty-two entries. The attractiveness of the Berlin contests for learned men all over Europe is illustrated by the nationalities of the laureates in the period from 1744 to 1786; twenty-six of these were German, eight French, two from Geneva, one Italian, and one from Transylvania. The international character of this competition was, in fact, guaranteed by the statutes adopted in 1744, which stated that "in cases where the treatises of a foreign and local scholar are matched as to thoroughness and beauty, the foreigner should always be given preference."[4]

While the prizes of the Berlin Academy, with the exception of that for philosophy, all concerned the mathematical or speculative sciences, those of other academies, among them the French Academy of Sciences, had a more practical orientation. The four prizes created by Louis XV and Louis XVI between 1766 and 1783, for example, were designed to encourage the industrial arts, notably methods for the exploitation of saltpeter and for converting salt to soda. Competitions in other academies concerned problems related to the practical arts of agriculture, animal husbandry, navigation, and manufacturing. This was the case in the Royal Swedish Academy of Sciences, founded in 1739, whose first prize, instituted in 1754, was designed to elicit useful findings in the areas of agriculture and husbandry.[5]

A greater concern with theoretical science is reflected in the new prizes that were created after the French Revolution when the French Academy of Sciences was reorganized, in 1795, as the First Class of the Institut National des Sciences et des Arts. In a decree issued the following year, the First Class was assigned the task of awarding two prizes in general science, the funds for which would be allocated by the state. It decided to offer only one, however, a gold medal, valued at 3,000 francs. It was converted, in 1803, into a cash prize for the same amount. This prize, the *grand prix des sciences mathématiques ou grand prix des sciences physiques*,[6] came to occupy a prominent position in the resurgence of research in physical science in Napoleonic France. The influence of Laplace led several of the prize competitions to reflect his program of research in the physics of short-range molecular forces and imponderable fluids.[7] Prize competitions were also used by the academy as a means to throw light on or to settle scientific controversies. The best-known case occurred in 1819 when the diffraction of light was the subject for the *grand prix des sciences mathématiques*. The prize was won by Fresnel, whose essay went counter to the established particle theory of light and advocated, instead, a wave theory. This has rightly been considered an important instance of the academy using its prestige to help bring about a shift of scientific paradigms.[8]

From the end of the Napoleonic era until about 1850, fourteen new prize funds were established by the French Academy of Sciences (representing a total sum of about 50,000 francs paid out in prizes awarded annually or at longer intervals), five of them by the same person, Baron de Montyon, who had already made important gifts to the academy under the *ancien régime*. In the absence of royal patronage, the chief patrons of science were the scientists themselves. Of the nine prizes created by individuals other than the Baron de Montyon, five resulted from legacies made by scientists, two of whom (Cuvier and Laplace) had been members of the academy.

Yet by mid-century the role of prizes in the scientific culture was probably less significant than it had been in the eighteenth century. By the early nineteenth century, in the opinion of Harnack, "the prize competitions were no longer favored by the [Berlin] Academy and lost the importance they had had in the eighteenth century." In a number of cases, inadequate treatment of the questions led to the prizes being reserved, "a strong indication that serious scientific study was on the wane in Germany"[9] or, to offer an alternative explanation, that standards were rising. In France, too, the importance of prize competitions declined. Of the fourteen new prizes created between 1815 and 1850, only two called for competitions. More important, there was a shift of emphasis in the opinions of the academy and the scientific community with respect to the use of prizes and prize funds. These were no longer regarded solely as a means to reward work already accomplished, frequently in response to a question. Instead, ways were sought in which they could provide scientists with the resources for work they were planning or already had under way. This shift from a system of prizes to one of subsidies (or, as it became later, one of grants) for research was prompted by changes in approaches to scientific work. The most important among these were (1) the emphasis on experimental method and a concomitant need for laboratory equipment and facilities and (2) the ambition to carry out direct observations, whether they be of distant planets or in faraway lands, observations that required instruments and funds for travel.

As the financial needs of the scientific enterprise grew, scientists came to regard both existing and prospective prize funds as a means to supplement the meager subsidies they were receiving from the state or the funds they could set aside for research from their own incomes. This way of thinking is illustrated by the following anecdote told about Jean-Baptiste Dumas, chemist, minister of public instruction, and later in his life (beginning 1868) permanent secretary of the French Academy of Sciences. In 1848, Dumas had been forced temporarily to close his laboratory at rue Cuvier in Paris for lack of funds. One day he was visited by a foreigner who demonstrated his willingness to help by placing a stack of bills on Dumas's desk. His name was Jecker, he said, and he had at one time followed Dumas's courses at the Sorbonne. To this he attributed his successful business ventures in the budding chemical industry. Dumas declined the offer of financial aid for himself but suggested that Jecker make a donation to the academy. When Jecker died in 1851, the academy found itself the recipient of a legacy of 200,000 francs. In accordance with Jecker's will, the proceeds were used for an annual prize of 10,000 francs, awarded for the purpose of promoting progress in the area of organic chemistry.[10]

While the use of prizes to help pay for the cost of research depended on the individual prizewinner's decision to employ the funds in this manner, a more fundamental shift from prizes to grants was anticipated by two developments, both of which permitted the academy to pay out subsidies for research more or less officially. In 1847, the first grant-giving fund at the academy was set up, the Fondation Tremont. The express purpose of the fund was to aid scientists or engineers to meet the expenses of experiments or other work carried out "to bring glory to France."[11]

More important, starting about 1830, the academy was able to use surplus income from the several legacies left to it by Baron de Montyon to give grants for research purposes. Such income was generated by the accumulation of capital from investments and through the prizes not being awarded for want of worthy candidates. Given that the Montyon awards concerned medicine and especially public health, a minor interest of the academy, one can understand why (to quote Crosland) there was "pressure from the majority to make the Montyon funds more generally available."[12] This was made possible through a royal ordinance (1829) justifying the use of excess income to pay for experiments, construction of apparatus, and acquisition of new books as necessary for the judging of the Montyon prizes. (This argument was also to figure importantly in the negotiations between the executors of the Nobel estate and the institutions designated to award his prizes.) By stretching this provision to the limits that the Ministry of Public Instruction, its supervisory institution, would allow, the academy was able to pay out grants – often benefitting its own members – for research that had little or no bearing on the Montyon prizes, and even to subsidize its own publication, the *Comptes Rendus Hebdomadaires*.[13] The precedents that these grants set with respect to the use of prize funds and the considerable freedom the academy enjoyed in paying them out were probably more important, however, than the actual sums involved.

PRIZES AND THE NEW PATRONS OF SCIENCE IN THE LATE NINETEENTH CENTURY

In the second half of the nineteenth century, there was a striking growth in the creation of prize funds, particularly at the French Academy of Sciences. This was mainly due to the emergence of a new class of patrons of science made up of wealthy industrialists and financiers. Alfred Nobel (1833–1896) was typical of this group, and the prizes he founded represent the culmination of a movement of patronage directed at the sciences. A measure of the significance of this movement is found in the growth of prize funds at the French Academy of Sciences.

In the decade 1901–1910, the accumulated benefits flowing from such funds rose to 1.5 million francs or an average of 150,000 francs per annum (see Table 1.1). These sums represented more than double that of the decade 1871–1880, when the funds distributed amounted to 650,000 francs. The growth in the overall funds distributed was matched by that of the individual prizes: the three *prix Lacaze* (1865), each of 10,000 francs to be awarded biennially to the authors whose work had contributed most to progress in physiology, physics, and chemistry; the *prix Alberto Levi* (1891), 50,000 francs to be awarded once and for all to the discoverers of a means of preventing or curing diphtheria; and the *prix Osiris* (1899), which, with its award of 100,000 francs, became the largest prize in the academy. This latter paled in comparison

Table 1.1 Funds distributed from prize funds and grant-giving funds of the French Academy of Sciences, 1850–1915 (in francs)

	1851–1860	1861–1870	1871–1880	1881–1890	1891–1900	1901–1910	1911–1915
Prize funds	329,728	466,701	648,570	1,113,680	1,283,780	1,504,880	800,560
Grant-giving funds	5,500	15,025	10,875	80,700	154,750	392,400	461,900
Total:	335,228	481,726	659,445	1,194,380	1,438,530	1,897,280	1,262,460

Sources: E. Maindron, Les fondations de prix à l'Académie des Sciences: les lauréats de l'Académie 1714–1880 (Paris, 1881); P. Gauja, Les fondations de l'Académie des Sciences (1881–1915) (Hendaye, France, 1917).

with the five Nobel prizes (1896), however, each of which amounted to 150,000 crowns (i.e., about 210,000 francs) when they were awarded for the first time in 1901.[14]

At the same time as it was growing quantitatively, the prize system of the French Academy of Sciences underwent qualitative change, the overall effect of which was to make it a more effective instrument for encouraging promising scientists in their work. This was possible because of more liberal formulations of the terms of the new bequests made to the academy. After 1860, these terms hardly ever called for prize competitions. Most prize funds were established to reward achievement or to stimulate work in the broad areas of *sciences physiques* or *sciences naturelles*. But even when narrowly restricted to a specific problem in a given discipline – in medicine, for instance, that of finding cures for diseases still re- garded as incurable (e.g., cholera and diphtheria) – the terms often allowed paying out part of the prize money to scientists whose work, although it may not have solved these problems, had at least contributed to their solution.[15] When the terms imposed were too restrictive in the opinion of the academy, it sometimes exer- cised its right to refuse the bequest.

The more liberal terms of the bequests gave the academy greater freedom in the use of the proceeds. It seized this opportunity to aid scientists whose work was considered promising. Since a large number of scientists could be considered deserving, the practice (unless prohibited by the donor) of dividing a given prize among several laureates was instituted. It also became common to reward scientists several times, thus making prize money a recurrent source of financial aid.

That the prize system functioned to encourage and aid scientists in the most active stages of their careers is shown by the data on the age of laureates in the disciplines of mathematics, astronomy, physics, and chemistry (see Table 1.2). The mean age of the lau- reates in these disciplines was around forty at the time of their first prize. In this respect, but also with regard to type of reward offered, prizes at the Paris Academy differed from the reward system of the Royal Society of London, which was built around medals having little monetary value and hence being primarily symbolic rewards. A study of the Royal medals between 1826 and 1914 shows that from 1855 onward, these were normally bestowed on older and more established men of science, the mean age of

Table 1.2 *Mean age of French Academy of Sciences prizewinners at the time of first award in selected disciplines, 1850–1915*

	Before 1850	1851–1860	1861–1870	1871–1880	1881–1890	1891–1900	1901–1910	1911–1915	All years	No information	Prize awarded posthumously	Total (N)
Mathematics	—	40.2	42.8	39.8	44	38.4	42	42.4	41.3	—	—	—
(N)	—	(5)	(9)	(13)	(21)	(20)	(19)	(5)	(92)	28	—	120
Astronomy	27.5[a]	31.5	45.5	41.8	40.8	43.1	47	43.7	41.8	—	—	—
(N)	(2)	(10)	(6)	(13)	(25)	(16)	(16)	(18)	(106)	26	1	133
Physics	—	—	—	43	37.7	42.3	43.5	46.2	43.3	—	—	—
(N)	—	—	—	(5)	(3)	(14)	(25)	(10)	(57)	16	—	73
Chemistry	36[a]	43.2	39.6	40.1	40.8	35.9	43.3	44.3	39.7	—	1	—
(N)	(1)	(5)	(10)	(12)	(23)	(23)	(18)	(6)	(98)	5	1	104

[a]Winners of two or more prizes having received the first of these before 1850.

Sources: J. C. Poggendorff, *Biographisch-literarisches Handwörterbuch zur Geschichte der exakten Wissenschaften*, (Leipzig and Berlin, 1898–1938), Vols. 3–6; Académie des Sciences, *Index biographique des membres et correspondants de l'Académie 1666–1939* (Paris, 1939).

recipients for this period being around fifty-five years.[16] The mean age of Nobel laureates in physics and chemistry during the period 1901–1915 was about fifty.

Paralleling the growth of prize funds was a substantial increase in the number and size of the grant-giving funds established at the French Academy of Sciences. From a modest beginning in the decade 1851–1860 when the academy only paid out 5,500 francs, in the years 1911–1915 the academy disbursed grants that amounted to over 450,000 francs for general research support, equipment, and travel (see Table 1.1). The size of the latter sum was due to the establishment of several large grant-giving funds that benefitted the academy. The most important of these were the Fondation Debrousse (1899) and the Fondation Loutreuil (1910) with capital of 1 million and 3.5 million francs, respectively.[17]

The increase in the grant-giving funds was in keeping with the academy's ambition to encourage donations that permitted it not only (as it had always done) to reward work already accomplished but, in the words of its permanent secretary Gaston Darboux speaking in 1911, "to stimulate, subsidize and encourage research."[18] Darboux linked his call for such donations to a program that foresaw more activist roles for the academy than had previously been the case. He felt that the academy should take the initiatives required by the constantly changing nature of scientific activity; for example, it should discover and encourage new talent and involve in its activities all scientists of proven merit, particularly those working alone.[19] Such programmatic statements were indicative both of the new demands that changes in the scientific enterprise were making on its oldest organization and of the belief of academy leaders in their ability to meet this challenge.

A comparison of the reward system of the French Academy of Sciences with those of its sister organizations in other countries reveals that scientific philanthropy in the form of prize funds and grant-giving funds was more advanced in France than elsewhere. Tabulations (carried out by Forman, Heilbron and Weart) of prizes *and* grants awarded for research in physics by leading European academies indicate that the amounts paid out by the Paris Academy in 1900 were roughly three times as large as those of either the Royal Society of London or the academies of Berlin and Göttingen combined. In the latter institutions, however, the ratio of prize money to grants was 1:10, whereas in the French Academy of Sciences this proportion was the reverse.[20]

The insignificance of prize funds both in the German academies and at the Royal Society was due to the particular way in which the reward system of each institution had developed in the nineteenth century. In the Berlin Academy, for example, prizes continued to be linked to competitions although these had lost their importance as a stimulus for scientific research. When, in the latter part of the nineteenth century, the Berlin Academy moved into the area of research support through grants, these did not grow out of prizes as was the case in France. Instead, the grant program (amounting to between 45,000 and 75,000 marks [i.e., about 55,000 and 93,000 francs, respectively] annually) depended largely on the annual subsidy the academy received from the Prussian government for the purpose of general research support. Only after the turn of the century was the government subsidy supplemented, in a significant manner, by grant-giving funds established through legacies or donations by private individuals, many of them wealthy industrialists. Among these funds were the Theodor Mommsen Stiftung, the Herman Vogel Stiftung, the Karl Güttler Stiftung, and the Albert Samson Stiftung, the last representing a capital of close to 1 million marks (or about 1,250,000 francs).[21]

In the Royal Society of London the reward system of the eighteenth and nineteenth centuries was built around medals rather than prizes, thus following the course on which it was set by the creation of the Copley medal in 1709. Being primarily symbolic rewards, these medals reflected, as did the noble titles conferred on scientists by the Crown, the belief quoted earlier that scientists would not be moved for the sake of lucre. While the Copley remained the most prestigious medal of the Royal Society, others were subsequently added: the Rumford, instituted in 1796 to reward "the most important discovery or useful improvement which shall be made ... during the preceding two years on heat or light"; the two Royal medals (1862) awarded for "scientific work of exceptional merit"; the Davy (1864) given out annually for the most important discovery in chemistry; and the Darwin (1885) in reward for work of distinction in the field in which Darwin had labored.[22]

As was the case in the Berlin Academy, the grant program of the Royal Society developed independently of the institution's traditional reward system and was mainly financed by government subsidies. A modest beginning of such a program was made in 1849 when an annual parliamentary grant-in-aid was voted to the society for the promotion of scientific research. Under the influence of the movement to endow research, the subsidy was increased in 1876 to £4,000 annually (i.e., about 100,000 francs): It was to be paid out in grants to help meet the expenses of individual scientists. The society did not strive to increase the subsidy, which remained at a modest level throughout the late nineteenth and early twentieth centuries.[23]

In contrast to the Berlin and Paris academies, there was no significant growth in the private grant-giving funds of the Royal Society around the turn of the century, the overall annual income from these – including the medal funds – amounting to about £1,300 (or 32,500 francs). That a belated change in the society's policy with respect to donations had occurred, however, is indicated by a resolution adopted in 1900 and subsequently inserted in the society's yearbook. It made reference to the several medals that the society had to award each year and went on to state that "it is neither to the advantage of the Society nor in the interests of the advancement of natural knowledge that this already long list should in future be added to and that, therefore, no further bequests to be awarded as prizes for past achievements should be accepted by the Society." By contrast, "funds which the Society is free to use for general purposes are very few indeed.... The President and the Council desire ... to make it generally known that while they will highly receive gifts to be applied for special objects or for the benefit of particular sciences ... they consider that, in view of the varying necessities of science, the most useful benefactions are those which are given to the Society in general terms for the advancement of natural knowledge."[24]

THE NOBEL PRIZES AND THE REWARD SYSTEM
OF SCIENCE AT THE TURN OF THE CENTURY

The foregoing describes the evolution of different kinds of rewards in the three major scientific societies in Europe over two centuries. Turning now to the functions of prizes in the scientific culture at the time Nobel made his donation, one finds several, partly overlapping roles.

These are most clearly in evidence in the reward system of the French Academy of Sciences where, at this particular juncture, prizes were important both as symbolic rewards, conferring honor and prestige on their recipients, and as a means of furthering the research activities of the scientists who chose to use the prize money in this manner (and there is evidence that many of them did).[25] This duality of roles constitutes the closest link between prizes in the French Academy of Sciences and the Nobel prizes. Although Nobel's will does not make any reference to the role he foresaw for his prizes, in statements to witnesses at the time of the signing of the will he stressed as his prime wish that the prizes be used to reward scientists whose work had shown promise. Awarding them a large sum of money would guarantee them the financial independence and the material means necessary to carry on their work.[26] In the statutes drawn up for the Nobel fund and in their actual awarding, however, his prizes came closer to being rewards for past achievement than assistance for the promising. This was probably inevitable in view of the emphasis put in the will on rewarding "discoveries or improvements," terms that suggest fulfillment rather than promise. As a result, the role of the prizes as symbolic rewards was enhanced and their material value – that is, their role as a form of financial aid – became a subsidiary factor in the prize awarding. In fact, the sheer size of the sum of money represented by the Nobel prizes probably precluded their being used as a form of research grant, for it seems unlikely that any scientific society would have taken on the responsibility of handing over the largest existing prize to scientists who had merely shown promise.

Another important link between the Nobel prizes and the prize system of the French Academy of Sciences is found in the role played by the relevant institutions in the awarding of prizes. In each case, it was the institution that translated the donors' wishes into practical terms, thus bringing these wishes into harmony with the scientific culture and the needs of science. In doing so, each institution made sure that its own needs and those of its members were not forgotten. It was shown in previous sections how the French Academy of Sciences derived material benefits from its prizes, both through paying out stipends to the judges of prize competitions – the *épices* in the will of Rouillé de Meslay – and from the grants (emanating from surplus income of the prize funds) that often benefitted members of the academy.[27] The Swedish institutions designated to award the Nobel prizes in science and medicine did not hesitate to claim such benefits for themselves, but here too their views were strongly influenced by prevailing opinions on the needs of scientific research. An important part of the negotiations carried out (1897–1900) between the executors to Nobel's estate and representatives of the four institutions designated to award his prizes concerned the attempt (ultimately successful) made by the Royal Swedish Academy of Sciences and the Karolinska Institute to use a portion of the fund to endow research institutes in the sciences (see Chapter 3). The establishment of such institutes – the future Nobel Institutes – must have struck the scientists negotiating the statutes as being a less anachronistic solution to the problem of making the prizes benefit the material needs of science than the doling out of subsidies under the guise of prizes. In this way, too, the role of the prizes as primarily symbolic rewards was enhanced.

In various provisions of Nobel's will there are echoes of the long history of prizes and medals sketched out above. One author has noted a marked similarity between certain expressions in the will and the regulations governing the Rumford medal, especially the provision that preference always be given "such discoveries as . . . tend most to promote the good of mankind."[28] These similarities were, however, chiefly those of language, the intentions behind the two awards – one symbolic and the other material – being very different. References to "the good of mankind," "human well-being," and "the public interest," not to mention "the glory of France," are also frequently found among the aims that donors to the French Academy of Sciences sought to promote through their prize funds.[29] The simultaneous endowment of prizes in five areas of human endeavor – in Nobel's will, physics, chemistry, medicine or physiology, literature, and peace – was a more original feature, for which there was nevertheless some precedent. In 1874, an Italian merchant, Jerôme Ponti, donated his entire fortune of 2 million francs to the Royal Society of London, the French Academy of Sciences, and the Academy of Sciences of Vienna. His will stipulated that the proceeds be used to institute annual prizes in the areas of mechanics, agriculture, physics, chemistry, travel by land or sea, and literature. This scheme was never put into practice, however, since the Ponti family succeeded in having the testament declared void after a lengthy court battle.[30]

As shown above, the Nobel prizes had their origin in national prize systems, the French one in particular. Their role as international awards, however, represented a significant break with past practices. Although foreigners were the beneficiaries both of the prizes of the French Academy of Sciences and of the medals of the Royal Society of London, there was no precedent in these or other scientific societies for paying out large sums of money to nationals of other countries. It was also very unusual for such societies to have foreigners participate in the decision making when prizes and medals were awarded. In both respects, the Nobel prizes were to act as precursors and emerge as the first truly international prize in modern times.

When the Nobel prizes were instituted, it was possible to speak of an international reward system of science built mainly on symbolic rather than material rewards. This system was much less clearly differentiated, however, with respect both to type of honors and to their prestige value, than that which functioned within national scientific communities. Scientific societies possessed relatively few awards that were regularly bestowed on foreigners; for this reason, perhaps, the distinction between prizes and medals did not apply internationally. The Copley, Rumford, and Davy medals of the Royal Society of London can thus be held to have ranked on a par with the *prix Jecker* and *prix Lacaze* of the French Academy of Sciences as major international honors.

Election to foreign membership of important scientific societies was another honor that often went hand in hand with the awarding of prizes and medals. There were also the honorary doctorates given out by universities to foreigners. The fact that these were available in large numbers may have deflated their value, but the worth of the honorary doctorates from a small group of prestigious institutions (the Sorbonne, the universities of Oxford and Cambridge, and the leading German universities) remained. It was not unusual for scientists to be awarded foreign decorations: the *Légion d'honneur*, in particular, was frequently used to reward foreign friends of French science and scientists. The plethora of honors that could be bestowed on a scientist of international renown is perhaps best illustrated by Wilhelm Röntgen who, between his discovery of X-rays in 1895 and his being awarded the first Nobel prize in 1901, was elected to membership of twenty-two German and foreign scientific societies, received seven major prizes and medals (among them the Rumford medal and the *prix Lacaze*), was made Königlichen Geheimrat by the government of Bavaria, and was honored with a statue in his likeness erected on the Potsdam Bridge in Berlin.[31]

As the example of Röntgen shows, the scientific culture around the turn of the century, reflecting the values of the societies in which it was embedded, was amply provided not only with occasions but also with the means to celebrate scientific achievement. The honors listed in the foregoing brought forth such celebrations, as did the anniversaries and jubilees of scientists, scientific congresses, international exhibitions, and the commemoration of the founding of important institutions. Extending these honors and celebrations to foreign scientists and scientific institutions provided tangible proof of the universalist creed to which most scientists subscribed – at least on such occasions. On the level of the day-to-day workings of science, these honors and the ceremonies that accompanied them served to reinforce the networks for communication and evaluation of research findings across national boundaries, most directly perhaps because these ceremonies provided an opportunity for scientists to meet.

Honors also played a role in furthering international scientific enterprises: The decorations solicited from the patrons of such enterprises – governments or reigning monarchs – could be used, for instance, to stimulate or reward participating scientists. The repeated proposals that the Nobel prize in physics be awarded to the directors of the International Bureau of Weights and Measures in Sèvres, and even to the bureau itself, can be viewed as attempts to use a high international honor to promote an international scientific enterprise that, furthermore, would have greatly benefitted from the money involved.[32]

The movement to reward scientists with cash awards has been described here primarily as the maturation of practices that had grown up inside scientific institutions. What should not be overlooked, however, is that this movement resulted as much, or more, from general socioeconomic and political conditions linked to industrialization that brought forth a new class of patrons of science made up of industrialists and financiers. Nobel was typical of this group, for like many donors of prize funds to the French Academy of Sciences at this time, he had experience with scientific work resulting in technological development. His donation also resembled many of the benefactions received by the academy in that it was not made with the intention of stimulating work of immediate practical value for industry – as were, for instance, the gifts that many industrialists made directly to scientific laboratories. Instead, these donations represented a way for industrialists with a philanthropic bent to give tangible proof of their belief in science as a moving force of human progress and to associate themselves, or their memory, with a prestigious social activity. By creating a prize, the donor could rest assured that his name would be perpetuated in the annals of science, and this regardless of the size of the bequest.

However much it was part of the mores of a new class of wealthy industrialists, for philanthropists to make a bequest on the scale of the Nobel fund – that is, of the order of 30 million crowns (or $8 million)– still required a combination of fortuitous circumstances. Whether or not the word "fortuitous" should be applied to the circumstances under which large bequests were most easily made (i.e., by those who had remained unmarried and childless) depends, of course, on one's perspective; while large donations brought joy to the institutions receiving them, they frequently provoked the ire of disinherited relatives. Be that as it may, the absence of claims on the estate by legal heirs, or the lack of success of such claims, represented a priori conditions for the realization of the testator's wishes. The case of the Jerôme Ponti fund mentioned above represents an example of the success of such claims; as will be described in Chapter 3, the claims advanced by the Nobel family against the fund were much less successful.

Nobel made only a few large donations during his lifetime.[33] In this he differed from other important contemporary philanthropists. In France, for example, there was Daniel Iffla, known by the name of Osiris, who had started as an accountant's helper in Bordeaux and made a fortune roughly equal to that of Nobel through astute investments and financial operations. Devoting the latter part of his life to philanthropy, he instituted the *prix Osiris* through a donation of 1 million francs. This was only one of many acts of largesse, not all of which benefitted the sciences, for he also made gifts with the aim of ameliorating the conditions of the sick and the aged, instituted *prix Osiris* in the amount of 50 francs each for pupils who had achieved excellence in the municipal schools of a number of French cities, and restored the château Malmaison before turning it over to the state. Iffla left no children or other close relatives, and the Institut Pasteur became his sole heir upon his death.[34] In America and the United Kingdom there was Andrew Carnegie, also a bachelor and probably the most remarkable philanthropist of all, who in the latter part of his life early in this century used a fortune many times larger than that of Nobel to create the ten Carnegie trusts, each designed to provide funds and set up institutions in given areas of social, cultural, educational, and scientific advancement.

That Nobel should have deprived himself of the pleasure of seeing his fortune put to use in encouraging and recompensing scientific achievement was due as much to personal predilection as it was to circumstances outside his control. Had he lived on instead of dying of a heart ailment at the age of sixty-three, he may well have engaged in active scientific philanthropy during his lifetime. However, his solitary nature and distrust of organizational efforts, manifested most clearly by his running his large business empire almost single-handedly, made it unlikely that he would have derived any satisfaction from using his fortune to participate in the setting up of new institutions or in expanding existing ones. When in his later years he made contacts with the scientific world outside the areas of his business interests, it was usually through correspondence with scientists and inventors who worked alone rather than in academic or other institutions.

It is not surprising, given Nobel's individualistic bent, that he should have chosen prizes as the chief form of his benefaction to science and culture, with all that these entailed of celebration of the lone explorer of mind and matter. But it also seems that he consciously sought to dissociate his life and person from the awards that he set up; for instance, there is no stipulation in the will that the fund or the prizes be named after him. By entrusting the awarding of the prizes to Swedish institutions, it almost seems as if he wanted to increase further the distance between the recipients of his benefactions and his own life, most of which had been spent either in St. Petersburg, where he moved from Sweden at the age of nine, or in Paris, which was his chief residence from the early 1870s until his death. Regardless of his reasons for this decision –

and many others have been suggested – it had a profound effect on the manner in which his wishes were transformed into concrete measures with respect to both the procedures for awarding the prizes and the selection of prizewinners. In order to learn more about the Swedish scientists and scientific institutions that were entrusted with the task of deciding, annually, on the "discovery," "invention," or "improvement" that "shall have conferred the greatest benefit on mankind" and hence most merited the prize, the next chapter will be devoted to a description of these.

2

Developments in Swedish and international science having a bearing on the Nobel institution

The state of Swedish science at the time of the founding of the Nobel prizes was governed as much by geography, demography, and tradition as it was by several new developments that had emerged in Sweden and elsewhere during the late nineteenth century. Located on the northern periphery of Europe, Sweden had a population of about five million and possessed a tradition of scientific achievement as well as a set of institutions – universities and scientific societies – capable of upholding this tradition and of carrying it forward. By the late nineteenth century, industrialization and modernization had provided a strong impetus for forward movement in higher education and research. The two universities in Uppsala and Lund, the country's traditional seats of learning, expanded in terms of both students and teaching positions. These developments particularly benefitted the physical sciences, which had been relegated to a secondary position since the early nineteenth century when idealism and romanticism, which put a premium on philosophy, classical studies, and esthetics, began to reign in the universities.[1] While the sciences were strengthened in the traditional universities toward the end of the nineteenth century, their importance was even more directly recognized at several newly created institutions, perhaps nowhere moreso than at the Stockholm Högskola,[2] which opened in 1878 as a private non-degree-granting institution primarily oriented toward the sciences.

Parallel with the strengthening of the institutional bases for scientific teaching and research, these activities themselves were undergoing change. Research came to play a more important role in the careers of university teachers as well as in the training of students. This, in turn, introduced the notion of specialization, a development that was tempered by two factors. On the one hand, the broad disciplines of physics and chemistry were represented by very few chairs, and so actual and aspiring chairholders were expected to master most of what was considered important in the subject. On the other hand, scientists were given (or seized) the opportunity to speak and write on many subjects other than their own research; like some of their German colleagues, they emerged as *Kulturträger* (upholders of culture). Generally seen as representing a new culture based on rationalistic and materialistic values, scientists gained in status and authority because of the many ways in which their knowledge was proving useful for ameliorating the material conditions of the country and its population. For most scientists, however, providing technical expertise for the expanding industries in the country – ironworks, hydroelectric power plants, pulp and paper mills, and so forth – was not as important as the new roles they assumed both in university reforms and, more generally, by infusing scientific knowledge and values into the cultural life of the nation.

It is not possible to understand the building of the Nobel institution without taking into account the reformist spirit in which a new generation of scientists had approached scientific institutions and activities in the 1880s and 1890s. The creation of new institutions was a characteristic of the times. It is not surprising, therefore, that among the men making important contributions when it came to bringing about Nobel's wishes were some of those – Svante Arrhenius (1859–1927) and Otto Pettersson (1848–1941),[3] for instance – who had forged the Stockholm Högskola into a successful enterprise. But the movement toward strengthening the scientific base of the country also carried over into the selection of Nobel prizewinners. Swedish scientists, members of the Nobel Committees for Physics and Chemistry, could take upon themselves the evaluation of proposals for the prizes because of prior developments that had resulted in both a differentiation of disciplines into specialities and research areas and an emphasis of experimentation and laboratory work. In both these respects, the evolution of Swedish science reflected trends that had already come to fruition in other countries, particularly Germany. It was natural, however, that in a small and peripheral country such as Sweden, scientific activities would to a large extent be shaped by local traditions and concerns. The research orientations represented among the members of the Nobel Committee for Physics, for example, owed much to the particular ways in which international trends had been adapted to local traditions: Conflicts between the Uppsala school of experimental physics and the much more eclectic approach of the physicists at the Högskola spilled over into the prize deliberations. At the same time, particularly in Uppsala, physics teaching and research had developed in a manner that left little room for mathematical and theoretical physics. Here was another element of local science that would influence the selection of Nobel prizewinners.

This chapter will focus on some specific features of Swedish science, institutional and intellectual, that have a bearing on the setting up of the Nobel institution and on the awarding of the prizes in physics and chemistry. The individuals who will figure importantly in subsequent chapters will be introduced. First, the different kinds of relationships that linked Swedish science and scientists to the centers of scientific activity, as well as to neighboring Nordic countries will be described. Second, some of the economic and cultural factors that contributed to the regeneration of Swedish science at the turn of the century will be briefly delineated. Third, pertinent aspects of the institutions that took on major responsibilities for awarding the science prizes will be analyzed. Chief among these institutions were the Royal Swedish Academy of Sciences as the final arbiter of the recommendations made by the Nobel Committees for Physics and Chemistry, and the science faculties in Uppsala and Stockholm as the bodies whose chairholders in physics and chemistry made up the bulk of the membership of these committees. (For a list of the scientists elected to each of these five-member committees from 1900 to 1915, see Appendix C.) Fourth and last, the main intellectual orientations of scientists in Uppsala and Stockholm will be the subject of attention. This chapter treats the period 1880–1915, particularly the years around 1900.

RELATIONS WITH THE CENTER AND THE
PERIPHERY

Sweden's small scientific community was quite naturally affected by its peripheral location with respect to the centers of scientific activity on the European continent and in England. Being on the periphery did not mean, however, that Swedish science was isolated from the influences of these centers. Given the country's long-standing cultural and linguistic ties with Germany, the influences that most broadly affected the form as well as the content of scientific production were those emanating from the German universities. In some specific fields, however, Swedish scientists maintained strong relations with other countries: for instance, France in mathematics.

In many ways, the relations of Sweden with the powerful scientific nations, particularly Germany, resemble the model outlined by Gizycki when he states that a peripheral community that perceives its relationship to the center as being competitive, at least in part, is neither overpowered by helplessness and hence reduced to isolation nor limited to an imitative response to the center. Rather, it selects, from the many types of influence radiated by the center, those that correspond to existing conditions or that can be adapted to these and regarded as improvements. The relation of the periphery to the center also undergoes change – slowly in the case of the process of emulation and adaptation but much more rapidly when the impetus comes from fortuitous circumstances, such as a major discovery or an institutional innovation.[4] Although, in the opinion of Gizycki, these are rare events in the periphery, they can nevertheless be held to have occurred in Sweden during the period studied here, one being Arrhenius's formulation of the theory of electrolytic dissociation and another the creation of the Nobel prizes.

That Swedish physicists and chemists were relatively close to the center is shown by the pattern of their training and publishing as well as by a more general awareness of research carried out abroad. Graduate or postdoctoral studies at one or several European universities, usually in Germany, were considered practically indispensable for those preparing for careers as academic scientists in the late nineteenth century. The *Wanderjahren* of Svante Arrhenius, for example, which extended from 1885 to 1890, took him to the universities of Riga, Würzburg, Graz, Amsterdam, and Leipzig, where he worked with, among others, W. Ostwald, L. Boltzmann, F. Kohlrausch and J. H. van't Hoff. To take another example, Knut Ångström (1857–1910), who was to become Arrhenius's colleague on the Nobel Committee for Physics, was apprenticed to A. A. Kundt, director of the Physical Institute of the University of Strassburg and one of the foremost German experimental physicists of the time. For these and other students of physics and chemistry the time spent at the large German universities possessing well-equipped laboratories was particularly important since it gave them an opportunity to learn new experimental techniques and to widen their practical experience of laboratory work.

The proximity to the center is also illustrated by the ease with which many Swedish scientists had their research results published in foreign journals, the ability to write German, French, or English being of course mandatory for academic scientists. Publishing in foreign journals was important not only because it meant that their research was judged by the international scientific community but also because there were limited opportunities for scientific publication in Sweden. In physics and chemistry, for example, there was only one specialized journal in Swedish, *Svensk Kemisk Tidskrift*, founded by the Swedish Society of Chemists in 1889. These and other contacts with the center had the effect of spurring an interest in scientific developments abroad, principally in the specialty areas where Swedish scientists were working but also more generally. A glance at the minutes of the physics societies (*fysiska sällskapen*) that Arrhenius helped organize, first in Uppsala (1886) and later in Stockholm (1891), along the lines of the German university colloquia, testifies to the speed with which the news of major scientific breakthroughs reached Sweden. For example, the results of work on cathode rays and radioactivity, later to be honored by Nobel prizes (e.g., those for P. Lenard, J. J. Thomson, and E. Rutherford), were all discussed at meetings of the physics societies in Uppsala or Stockholm very shortly after they had been announced.[5] (For the list of Nobel laureates in physics and chemistry, 1901–1915, see Appendix D.)

As these examples show, most Swedish scientists had some contact with scientific centers on the Continent or in England. Only for the very few, however, did such contacts translate into reputation and influence outside Sweden. Among these, Arrhenius and Gösta Mittag-Leffler (1846–1927) are the most important here since in both cases the positions they occupied in international networks directly influenced Nobel prize deliberations. How this occurred will be discussed at some length in Chapter 5.

Arrhenius's international position rested on his role as one of the founders of the research school of physical chemistry, whose members, collectively known as the Ionists, revolutionized the study of chemical processes through the application of thermo-dynamics. Arrhenius's central position in this network arose from the bold manner in which his 1887 theory, concerning the permanent dissociation of molecules into oppositely charged ions in extremely dilute solutions of weak electrolytes, brought together several strands of research. It integrated chemical studies of the nature of solutions and the phenomena of dissociation, the latter having been studied since mid-century through the vapor densities of heated substances, with the work of those physicists who had turned their attention to electrolytic phenomena conceived in terms of the interaction of electrically charged particles. The founding members of the school – Arrhenius, van't Hoff and Ostwald – were first brought together in the 1880s through the affinities they saw in each other's work. During his *Wanderjahren* on the Continent, Arrhenius became a personal friend of the other two and also established contacts with other scientists working in physical chemistry – among them, Planck and Nernst. Throughout the 1890s, it was Ostwald, holding the first chair in physical chemistry at the University of Leipzig, who took the lead in exploring the ramifications of Arrhenius's theory. The principal vehicle for the diffusion of the work of the Ionists was the journal *Zeitschrift für physikalische Chemie*, founded by Ostwald in 1887. Bridgeheads for the school were also created in England, where W. Ramsay was an early supporter of Arrhenius's theory, and in the United States through American students trained by Ostwald.[6] When Arrhenius returned to Sweden in 1891 as lecturer and later (1895) professor of physics at the Högskola, he was sufficiently well known internationally to attract postdoctoral students from abroad; in the years 1891–1900, some fifteen in all, an unprecedented number for Sweden, came to work under him. The large majority were Germans who had earned doctorates from prestigious universities, but there was also a sprinkling of Britons and Americans.[7]

In contrast to Arrhenius, Mittag-Leffler's international position rested less on his contribution to science, in his case pure mathematics, than on the determined way in which he had established himself as a central figure in a network of European mathematicians. When called to the Högskola in 1881 as one of its first professors, Mittag-Leffler was already well known on the Continent, having studied with K. Weierstrass in Berlin and C. Hermite and J. Liouville in Paris. In 1882, he founded *Acta Mathematica*, which under his editorship became an important outlet for publication by well-known European mathematicians. Mittag-Leffler was particularly close to the younger generation of French mathematicians whose leading members – G. Darboux, P. Painlevé, and H. Poincaré – all made important contributions to *Acta Mathematica*. Through their prominent positions in French academic life and in the French Academy of Sciences, these men represented an important scientific interest group, the influence of which extended to Nobel prize selections. Particularly in the early years of the Högskola, Mittag-Leffler was instrumental in bringing scientific talent to the fledgling institution. The most prominent example was the Russian mathematician Sonya Kovalesky, a student and collaborator of Weierstrass. She was professor at the Högskola from 1884 until her early death in 1891. As rector of the Högskola (1891–1892), Mittag-Leffler also helped provide teaching positions for Arrhenius and for Vilhelm Bjerknes (1862–1951), the Norwegian mathematical physicist. Among the many benefits that Swedish mathematicians gained from Mittag-Leffler's wide-ranging contacts were visits from foreign mathematicians; in 1895, Painlevé spent a semester as visiting professor at the Högskola, and in 1896 the same position was held by V. Volterra from Italy. When Mittag-Leffler celebrated his fiftieth birthday, four hundred colleagues from all over the world contributed funds for a life-size portrait, a tribute that was also a measure of the name he had made for himself in the world of mathematics.[8]

If close relations with the center often depended on the networks of particular individuals, those within the periphery – that is, among the Nordic countries – were much wider-ranging and of longer standing. Many students from the Swedish-speaking regions in Finland traditionally enrolled at the University of Uppsala. In the 1880s and 1890s, improved conditions of study and research in science and medicine attracted to Sweden scholars from all the Nordic countries, students as well as those who had already finished their training and were seeking academic employment. Superior training facilities, particularly in medicine, have been cited as an explanation for the increase of students from other Nordic countries both at the universities and at the Karolinska Institute, the medical faculty in Stockholm.[9] The opportunities for academic employment that arose from the creation of the Stockholm Högskola also launched the careers of several scientists who later became prominent, among others W. C. Brögger, V. Bjerknes, and N. Wille, all of who came to Stockholm from Oslo, E. Warming from Copenhagen, and Mittag-Leffler from Helsinki.

SCIENCE IN THE SERVICE OF PROGRESS

While the relationship of Swedish to international science is important in the context of the Nobel prize institution, it nevertheless affected only a part of the national scientific enterprise. In fact, it is possible, as illustrated by a recent study, to draw a picture of Swedish science in the late nineteenth century as an enterprise dominated by national interests and concerns.[10] In this description, mapping and inventory work are found to have played a predominant role in Swedish scientific life at the time. It is seen to have resulted from the way in which the traditional interest in natural history, which can be traced back to Linnaeus, was regenerated by the movement toward modernization and industrialization of the country. The disciplines that had traditionally occupied a strong position (e.g., botany, zoology, geology, and mineralogy) were most affected by this movement. By contrast, physics and chemistry were less influenced by practical concerns. That this nevertheless occurred is illustrated by the incorporation of physical principles in meteorology and the way in which chemistry spawned such important new subareas as agricultural chemistry, wood chemistry, and metallurgy.

The new emphasis placed on science in the economic and cultural life of the nation can be measured by the increase in the number of teaching positions both at the traditional universities and at the institutions (e.g., the Stockholm Högskola and the two institutes of technology) that offered courses comparable to those of the universities. In the years 1890 to 1914, the number of professorships grew from nine to seventeen in physics and from sixteen to nineteen in chemistry. Younger scientists were also offered employment opportunities through the strengthening of teaching and research staffs at lower levels. This came about partly through the creation of the new positions of *laborator* (associate professor), with special responsibility for supervising the laboratory exercises of students, and partly through the expansion of the category of *docent* (assistant professor). This latter position provided postdoctoral students with salaries for a period up to six years, thus permitting them to do research that would qualify them for higher teaching positions.[11]

Traditionally, the state had assumed responsibility for higher education; hence, the growth in scientific employment and facilities was mostly the result of increased government investments. This did not make for greatly increased government direction and control, however, since these investments were often channeled through institutions, the universities in particular, which, although state-supported, constituted largely autonomous corporations at least as far as appointments and programs were concerned. At the same time, the unversities were obliged to adhere strictly to the rules governing public institutions that had evolved through the long tradition of a strong central bureaucracy regulating all matters falling within the purview of the state. Under these rules, the keeping of records and accounts had to follow prescribed procedure, and all decisions of importance had to be confirmed by a supervisory authority (in the case of the universities, the Ministry of Ecclesiastical Affairs). Professorial appointments, which were always made by the king-in-council, were governed by a complicated set of special rules (the so-called *sakkunnighetsinstitutionen*), adopted in 1876. These involved, among other things, the appointment of independent experts (*sakkunniga*) who submitted detailed reports on the candidates and ranked them according to their qualifications for the chair in question. These are all matters that explain, at least in part, the detailed rules that were drawn up for the awarding of the Nobel prizes and the strict adherence to these rules in the actual process of prize selection.

The expansion of the scientific enterprise was sustained by several strands of ideology, chief among which were nationalism and a form of Darwinian evolutionism that stressed the progress of societies and their populations toward ever higher levels of "civilization," thereby meaning both material and intellectual attainments. The interaction between these two sets of values fostered the view that progress depended upon the "peaceful contest" among nations and that nations could also be ranked according to their level of "civilization." Scientific achievements were part of a high level of "civilization" since they could be shown to affect the material as well as the intellectual well-being of the nation. When these also had strong popular appeal, as in the case of the great

Arctic expeditions, particularly that of A. E. Nordenskiöld in 1878–1879, which led to the opening of the Northeast Passage, they fostered a sense of national identity and became occasions for celebrating the virtue of Swedish scientists and, indirectly, the Swedish nation.

Perhaps because the values that found expression in nationalism and evolutionism were so broadly defined and malleable, they were widely shared among the country's small elite – higher civil servants, university professors, and representatives of industry. This can be exemplified by the way these values were brought into play in the setting up of the Nobel institution as well as by the broad support that the institution received from the government and, after some initial hesitation, from the country's scientific and cultural elite. For those who subscribed to nationalist values, the fact that Swedish institutions had been chosen as the adjudicators of what was seen as a worldwide contest among nations was a source of pride and a challenge for the nation to acquit itself of this task in an honorable manner. As a "peaceful contest" among nations, the prizes also fitted the evolutionist view, but here they were invested with a supranational purpose since the progress that would ensue from such a contest was seen as eventually benefitting all nations. The provision in Nobel's will that the discoveries rewarded should be those conferring "greatest benefit on mankind" was in harmony with the belief that science and culture transcended national boundaries; this helped those who awarded the prizes to view their work in international perspective.

Of these strands of ideology, a form of evolutionism termed active or "optimistic evolutionism" had the most direct influence on scientists' conceptions of their disciplinary and cultural roles. Although scientists did not reject nationalist values, many of those who held liberal political views associated nationalism with conservatism and found that it had overtones of both chauvinism and militarism, as was the case, for example, at the time of the crisis over Sweden's union with Norway.[12] Optimistic evolutionism was represented in most industralized countries, yet it was never formalized into any one school of thought. It was expressed in a number of different ways: in the works of philosophers such as H. Spencer and H. Höffding; by the monists, led by the German biologist E. Haeckel, for whom science amounted to a new religion; and in the popular writings of T. Huxley, Ostwald, and

Arrhenius. Arrhenius believed that the contribution of science to progress was not a matter of *Weltanschauung* but of practice, and he therefore stressed the many ways in which science could enhance the adaptation of man to his natural environment and bring about greatly improved living conditions: the increases in agricultural production engendered by hybridization and selective breeding, for instance, or the more effective use of natural resources that would result from the introduction of science-based technologies.[13]

Like many of his colleagues, Arrhenius felt that putting science to work for the betterment of the world was an ideal to be pursued actively. For Swedish scientists who acted on this premise, however, applied research or specialist roles as consultants to industry or the government were not as important as their activities in bringing the results of scientific research to the attention of the public through popular lectures or articles in the press. The importance that many scientists in Sweden and elsewhere attached to popularization is another example of how the Nobel prizes fitted the scientific culture of the time: That the prizes constituted a particularly powerful medium for propagating and propagandizing science would become apparent soon after they had begun to be awarded.

In Sweden, the flourishing of popular science was helped by two conditions: on the one hand, the interest of the public, which was partly a function of the pride the nation took in the achievements of its scientists, and on the other, the way in which the values of optimistic evolutionism were translated into action within the framework of the liberal reformist movement. To raise the level of education in the country and to combat all forms of ignorance – whether they be those that caused dangers to public health through unsanitary living conditions or to individual self-realization through prejudice and "old-fashioned" thinking in the areas of religion or matrimonial relations, for example – were important goals of the liberal student association Verdandi, founded in Uppsala in 1882. It was also through Verdandi and similar associations that scientists with a reformist bent were provided with an outlet for their popular writings; the pamphlets of Verdandi were widely distributed and covered a broad range of topics relating to scientific research and social policy. As such they reflected the many different meanings attached to the term *culture* in liberal opinion; these were also the bases on which scientists emerged as *Kulturträger*. Being more concerned with culture than with politics, most scientists kept their distance from political parties, particularly the Social Democratic movement.[14] By maintaining a certain neutrality in this respect, they could safeguard their reputation as bearers of objective knowledge in the eyes of the public and also be part of a national elite that included many conservatives with little sympathy for opposition movements.

Vilhelm Carlheim-Gyllensköld (1859–1934), member of the Nobel Committee for Physics for almost twenty-five years, provides an example of the multiplicity of interests and roles that characterized the careers of some Swedish scientists at the turn of the century. A contemporary of Arrhenius during his undergraduate days at Uppsala, Carlheim-Gyllensköld remained at the university throughout the turbulent 1880s and was one of the members of Verdandi whose outspoken attacks on "bourgeois morality" scandalized the university establishment. At this time, he was also trying his hand at drama; when he sent a play to Strindberg, the latter felt it foretold a great dramatist. This was not to be, but the two remained in close contact, as evidenced by Strindberg naming Carlheim-Gyllensköld his literary executor.[15] In physics, Carlheim-Gyllensköld's work centered on geodesy, particularly terrestrial magnetism, where his skills in mathematics led him to advance some innovative formulations concerning the magnetic potential of the earth's crust at different historic periods. Like many of his contemporaries, Carlheim-Gyllensköld applied his knowledge in practical and popularizing pursuits. To cite only one example, in 1898 he participated in the large Swedish–Russian Geodetic Survey on Spitzbergen and wrote a popular account, *På åttionde breddgraden* (Along the eightieth parallel), which reflected his interests in the geophysics of the Arctic, particularly the aurora borealis or northern lights. Professor of physics at the Högskola from 1907 onward (but after 1911 in name only), he lectured on different topics of general physics, mostly those bordering on mathematical and theoretical physics. These latter areas became the foci of the evaluations he carried out for the Nobel Committee of Physics, which he joined in 1910. Even before joining the committee, however, he benefitted from the warm welcome that leading European physicists extended to Swedish scientists, often as a result of Nobel prize attributions. In 1907 he worked under J.

J. Thomson at the Cavendish Laboratory in Cambridge, and in 1908 he spent some time with H. A. Lorentz in Leiden.[16] In the latter part of his life, he devoted much energy to the creation of a museum for the history of science at the Academy of Sciences. Summing up his life's work, one of his biographers described it as being perhaps most akin to that of a natural scientist in the eighteenth century, thereby providing a reminder of the fact that, although part of the modernization of the country, the science practiced at the turn of the century also had strong roots in the past.[17]

INSTITUTIONAL TRADITION AND INNOVATION

The steady expansion of science and science-related teaching and research that had begun around 1890 and continued up until the First World War affected not only traditional institutions but, to an even greater extent, new ones such as the Stockholm Högskola. Among the former, the Royal Academy of Sciences is of particular interest here. To understand the role that the Academy came to play in the selection of Nobel prizewinners in the sciences, it is necessary to describe the morphology of this institution. The science faculties in Uppsala and Stockholm will then be described, and I shall focus on the departments of physics and chemistry, which provided the majority of the members of the Nobel committees in these disciplines.

The Royal Academy of Sciences

The country's oldest corporation of scientists, the Academy had as its most important function the offering of recognition and reward for scientific achievement – first, by electing scientists to membership, and second, by publishing research and awarding prizes and research grants.[18] The Academy's *Proceedings* (*Handlingar*) and journals were major platforms for the publication of research by the country's scientists. The prizes and grants of the Academy, though, were neither numerous nor lavish, especially when compared with those of the continental academies: The largest grant by far was the Letterstedt travel stipend (4,000 crowns).[19]

In addition to these traditional functions, the Academy was responsible for a range of tasks in the area of science administration, chiefly those relating to its own institutions. The most important institution, placed under the supervision of the Academy but financed by the state, was the Museum of Natural History, with its six departments each headed by a scientist (A. E. Nordenskiöld being one of them) holding the title of professor. Through its Bureau of Meteorology, the Academy had responsibility for the nation's weather service. The Academy was also in charge of the Astronomical Observatory in Stockholm. In physics and chemistry, however, the Academy's own institutions had fallen into abeyance. The Institute of Chemistry, founded in 1849, had functioned only for a few years. The Institute of Physics, founded at the same time, was still in existence at the turn of the century, but it functioned on a very modest scale. It was staffed by the Academy physicist Bernard Hasselberg (1848–1922), who by virtue of this position became the first chairman of the Nobel Committee for Physics.

Traditionally, the Academy had emphasized natural history, medicine, and the practical arts of agriculture, husbandry, and ironwork. This was reflected in the different sections (*klasser*) into which the Academy was organized. While the preponderance of the practical arts had diminished through the reorganization instituted in 1821 by Berzelius, as Academy secretary, the experimental sciences were still largely overshadowed by natural history and medicine. At the turn of the century, the members of the Academy (one hundred in all) were divided into nine sections, the largest ones being zoology and botany, medicine and surgery, economic sciences, and knowledge in general, all of which had either fifteen or sixteen members. By contrast, the physics section only had six members, and chemistry combined with mineralogy had twelve. This organization remained in force until 1904, when the task of selecting Nobel prizewinners in physics and chemistry made the Academy aware of the need to strengthen the two sections having primary responsibility for the prizes. As a result, the section of physics (combined with meteorology) was increased to ten members, and chemistry was given its own section, also with ten members.

Given the limited number of chairs available, only a segment of the community of physicists and chemists could be represented at the Academy. In 1900, the physics section was made up of B. Hasselberg, the Academy physicist; two meteorologists (R. Rubenson and H. H. Hildebrandsson); the former and present holders of the chair of physics at Uppsala (R. Thalén and K. Ångström); and the old Baron Fock, a former professor of physics at the Stockholm Institute of Technology and a member of Parliament. One has to agree with Bjerknes when he states that "it was very much an open question whether real competence in physics resided inside or outside the Academy when it took on responsibility for awarding the Nobel prizes."[20] In Bjerknes's opinion the latter situation predominated, and he goes on to cite the names of prominent physicists who did not hold chairs at the Academy.

First and foremost, there was Arrhenius, whose reputation in the field of physical chemistry was confirmed by the mid-1890s – if not at home, then certainly abroad – but who was not elected to the Academy until 1901, and then only with difficulty.[21] Another name mentioned by Bjerknes is Ivar Fredholm (1866–1927), "without doubt Sweden's leading mathematician" and one of the few Swedish scientists who was well versed in mathematical physics. When Fredholm was finally elected to the Academy in 1914, he was given a chair in the section of applied mathematics rather than in pure mathematics or physics. Bjerknes does not argue his own case, which was certainly a valid one, but mentions only that he, like many other Norwegians, was not elected to the Academy at this time. Although, in principle, Norwegians had been eligible since the time the two countries entered the Union, in fact, only four were elected between 1880 and 1900. On the whole, the chemistry section, which was double the size of the physics section until the reform of 1904, was more representative of the national community. It included the chairholders in chemistry at the universities of Uppsala, Lund, and Stockholm (O. Widman, P. T. Cleve, C. W. Blomstrand, and O. Pettersson) as well as the two agricultural chemists (L. F. Nilson and H. G. Söderbaum) who successively directed the research station of the Academy of Agriculture in Stockholm.

The Academy also elected foreigners as members (seventy-five in all), who were organized in sections paralleling those of the Swedish membership. Being mostly honorary, membership in the foreign sections was part of the system of exchange of honors that linked the different academies of science in Europe. All of the French physicists and chemists elected to the Academy, for instance, were long-standing members of the French Academy of

Sciences (A.-H. Becquerel, M. Berthelot, A. Cornu, C. Friedel, E. Mascart). Foreign members were to become important as nominators for the Nobel prizes in physics and chemistry; before then the relationship was sometimes so distant that the Academy failed to take note of the death of one of its members.

Still, some foreign members maintained relationships with the Academy that were beneficial for both parties. When Alfred Nobel was elected to the section of economic sciences in 1884, he wrote to his sponsor, A. E. Nordenskiöld, whose Arctic expeditions he had helped finance: "I consider the honor given me by you and your colleagues not as a reward for the little I have been able to accomplish but as an encouragement for future work. If with such a stimulus I do not manage to do something useful in the area of progress, I shall bury my wretched soul alive in some forgotten part of the world."[22] At the time, it would have been difficult to imagine, however, that the Academy was to be as handsomely rewarded as it was in Nobel's will.

For a scientist such as van't Hoff, foreign membership was linked to the publication of his treatise "Lois de l'équilibre chimique de l'état dilué, gaseux, ou dissous" in the Academy *Proceedings* (1886). The treatise had been submitted to the Academy by Pettersson, who was at that time *docent* in chemistry at Uppsala University and who had provided van't Hoff with data on the latent heat of fusion. In this work, van't Hoff showed, as a strong probability, that Avogadro's law was valid for dilute solutions, a finding that was to figure importantly in Arrhenius's formulation of his theory of electrolytic dissociation.[23] Van't Hoff was made a foreign member in 1892 and received the first Nobel prize in chemistry in 1901. He was not the only foreign member so honored, since of the twelve foreign scientists making up the chemistry section in the 1890s, four subsequently became Nobel prizewinners (van't Hoff, von Baeyer, Lord Rayleigh, and Ramsay).

As the official body representing Swedish science, the Academy also maintained contact with international scientific organizations. Of these, the International Committee on Weights and Measures was to figure as an important listening post abroad for some members of the Nobel Committee for Physics. The International Committee was set up to promote the metric system and to oversee the research work in metrology carried out at the International Bureau of Weights and Measures, headquartered in Sèvres, outside

Paris. The Swedish seat on the committee was occupied first by Thalén and, from 1900 onward, by Hasselberg. Both men had a strong interest in metrology, which stemmed from their own research specialty, spectroscopy. Extremely precise measurements were required to map out the spectra of different elements and to compare these against a standard scale based on the wavelengths of the solar spectrum.[24] The Academy was also a member of the International Association of Academies created in 1899, and it sent representatives to the large meetings of the association held in Paris (1901), London (1904), and Vienna (1907).

The science faculty at the University of Uppsala

The strengthening of the science faculty at Uppsala from the 1870s onward provided for a stronger position of the sciences in the university organization. The more specialized curricula that were instituted stressed research and laboratory training as an important part of general science education (mainly oriented toward producing schoolteachers) and as an essential qualification for academic positions. The first of these developments occurred in 1876 when the faculty of philosophy was split into two sections, one for the humanities and the other for mathematics and the natural sciences. In terms of the number of chairs, though, the former was to remain the predominant one until the 1930s. The second development was more gradual. The introduction in 1870 of the *filosofie licentiat* as a degree preparatory to the doctorate introduced the notion of specialization in graduate study, but it was not until 1907 that the three subjects required for the *licentiat* were reduced to one. On the undergraduate level, the abolition of the obligatory study of Latin and the successive reduction of the number of subjects required for the basic degree (*filosofie kandidat*) were steps in the same direction. A more important development perhaps was the introduction of set curricula for the different subjects, which, as one observer has put it, ended the students' "wanderings around the sea of knowledge" and also substantially reduced the chairholder's prerogative to determine the content of his subject, at least for the levels below the doctorate.[25]

The formal curricula instituted in physics and chemistry around the turn of the century required students to master the major subareas of these disciplines, acquiring much of this knowledge

through lectures and encyclopedic textbooks; in physics, Warburg's *Experimentalphysik* was the prescribed text. Practical exercises in the laboratory were an integral part of the curricula. This
was less of a novelty in chemistry, which traditionally had been
taught in this manner, than in physics, where the German model
was now followed by teaching students experimental skills in the
laboratory. Here they were guided by Kohlrausch's *Leitfaden der
praktischen Physik*, which for each major physical phenomenon
(heat, light, electricity, magnetism, etc.) described the appropriate
instrumentation and gave instructions as to how it should be used
to arrive at precise measurements. The emphasis placed on laboratory work made it necessary to improve existing facilities for
research. In the early years of the century, physics and chemistry
had outgrown their shared building, which dated back to the 1860s.
Separate new institutes were established, chemistry in 1904 and
physics in 1908. While the building programs provided both departments with adequate space, equipment and materials for research presented a more serious problem due to insufficient funds.
At the turn of the century, the annual budget of the physics laboratory (not counting salaries) was around 3,000 crowns, and it
had grown very little since the 1880s.[26]

The growth of teaching staffs that occurred in both physics and
chemistry during this period resulted from the differentiation of
these fields into specialties represented by their own chairs. This
specialization proceeded along somewhat different lines in physics
than in chemistry, something that was to influence the composition
of the Nobel committees in each of these disciplines. Traditionally,
the holder of the chair of general physics (who was also the director
of the Institute of Physics) represented the experimental orientation
that had dominated physics research and teaching since the middle
of the nineteenth century. The research orientations of Thalén and
Ångström, who occupied the chair from about 1870 to 1910, were
further specialized to spectroscopy and other radiation
measurements.

Following developments in Germany, where mathematical
physics emerged as a separate discipline in the last third of the
nineteenth century, a chair was instituted outside Uppsala's physics department in the 1870s. The choice of G. Lundquist (1841–
1917), who held the chair for some thirty years, seems to have
been an unfortunate one, however, and his teaching failed to take

account of new developments in mathematical physics and was carried out largely in isolation from the physics department, perhaps because of opposition from members of the latter.[27] This situation did not change until C. W. Oseen (1879–1944), Lundquist's successor in 1911, undertook a determined effort to introduce theoretical physics into Uppsala and, after the First World War, into the Nobel Committee for Physics as well.[28] The creation of a chair in meteorology (occupied by H. H. Hildebrandsson) at about the same time as that of mathematical physics was another instance of differentiation that took place outside the physics department. Thus throughout the period studied here, the latter remained a stronghold of experimental physics, an orientation that was carried over to Nobel prize selections through the Uppsala physicists who served on the committee.

In chemistry, by contrast, the chairs that resulted from the introduction of new specialties were created within the department of chemistry or in close cooperation with it. While the main chairholder, P. T. Cleve (1840–1905), was mainly interested in inorganic chemistry, O. Widman (1852–1930), who occupied an *extraordinarius* professorship in the department, was an organic chemist. Working in close cooperation with the chemistry department was O. Hammarsten (1841–1932), who occupied the chair of medical and physiological chemistry from 1883 onward. In the case of both men, the impetus for the introduction of their specialties into Sweden is said to have come from their close contacts with continental research schools: Widman with the school of A. von Baeyer in Munich and Hammarsten with that of C. Ludwig in Leipzig.[29] Cleve, Widman, and Hammarsten all served on the Nobel Committee for Chemistry, ensuring that a range of specialties were represented on the committee, which was also reflected in the prize decisions.

The narrow focus on spectroscopy in the physics department strongly influenced research training for higher degrees, since in the absence of paid laboratory assistance, a professor relied heavily on students to advance his work. During the twenty years of Thalén's professorship graduate students learned, as Arrhenius was to put it later, that "spectral analysis was considered the only part of physics worth pursuing."[30] When Arrhenius sought to carry out the laboratory work required for the *licentiat* in electrochemistry, he was denied access to the laboratory on the grounds that there was no space and that, in any case, chemistry and not physics should be his area of specialization. Although Arrhenius's case was probably an extreme one, the hierarchy entrenched in the Uppsala science departments coupled with limited opportunities for recruitment and advancement made the Stockholm Högskola attractive to many younger scientists, among them Arrhenius himself, Carlheim-Gyllensköld, and Pettersson.

The Stockholm Högskola

From its foundation in 1878, the Högskola had been characterized by its supporters as the antidote to the universities both in terms of its aims and organization. It had neither a formal entrance requirement – the school being open to any man or woman having some grounding in one of the subjects taught – nor did it issue formal degrees in these subjects. Conferring degrees recognized by the state was, of course, a right reserved for the universities and was not sought by the Högskola during the first twenty years of its existence. The early aims and ambitions of the institution were eloquently described by Pettersson writing to Ångström in 1883: "Anyone with scientific aptitudes and interest is welcome to us. He has the right to fix upon the subject of his choice and, at any stage, get a certificate concerning his ability.... You can rest assured that our certificates will be respected. This is in the interests of the students and to safeguard that [interest] *is* the overriding aim of the Högskola. The universities will be taught a lesson when they see that one can prove one's skills and knowledge ... without submitting oneself to the crazy tests ... required for the *kandidat*, etc."[31]

The Högskola was also unusual in its organizational structure, which represented a conscious break with university tradition. The day-to-day care of the institution was placed in the hands of a teachers' council. There was also a governing board composed of representatives of cultural and scientific institutions. The rector, *primus inter parens*, was elected by the council, whereas the governing board had final say in the appointment of teachers. The prerogatives and duties of the council and the board having been left rather vague in the statutes adopted in 1877, these matters were to be a source of considerable conflict during the early years.[32] This was probably inevitable given the emphasis put on democratic decision making by members of the teachers' council as well as the general tendency to do away with hierarchical structure. Innovation was the order of the day even on the semantic front, where the Latinized terminology in use at the universities (*consistorium, rector magnificus*, etc.) was replaced by simpler Swedish terms. For the men who were to make innovative contributions to the setting up of the Nobel institution, the experience they had gained in confronting novel situations and problems at the Högskola doubtless stood them in good stead.

At the turn of the century, the quality of teaching and research in many fields of science at the Högskola was probably on a par with, or even superior to, that of the science faculties at Uppsala and Lund, although quantitatively the institution was still small, with only ten professors (all except one in the physical or natural sciences) and few students. Whereas in its early years the content of the lectures was perhaps more popular than scientific, full-fledged courses were later instituted, with laboratory training a vital part of the teaching, particularly in physics and chemistry. At this time, the proportion of students who took courses and used the laboratory facilities at the Högskola to prepare for the degrees of *filosofie kandidat* or *filosofie licentiat* at one of the state universities had also increased significantly. In terms of both finances and facilities, however, the institution continued to lead a precarious existence, being wholly dependent on donations from foundations and wealthy individuals. In the late 1890s, the Högskola's physics institute received an annual grant of 1,600 crowns from the Högskola, a sum that represented about half that allocated at the state universities. The modest scale on which the physics laboratory was equipped is illustrated by the total worth of its apparatus, estimated in 1898 at 27,000 crowns.[33] A more serious shortcoming was the lack of space. Until a specially constructed building was erected in 1909, the Högskola was housed in various rented facilities. Hans von Euler (1873–1964), the German physical chemist who came to work with Arrhenius in 1897, later recalled his "surprise and amazement, having just come from Nernst's institute in Göttingen, in seeing these rooms [the physics institute] and the apparatus in them."[34]

The departments of mathematics and the physical sciences at the Högskola were more attuned to new developments abroad than were those in Uppsala, mainly because of the wider range of

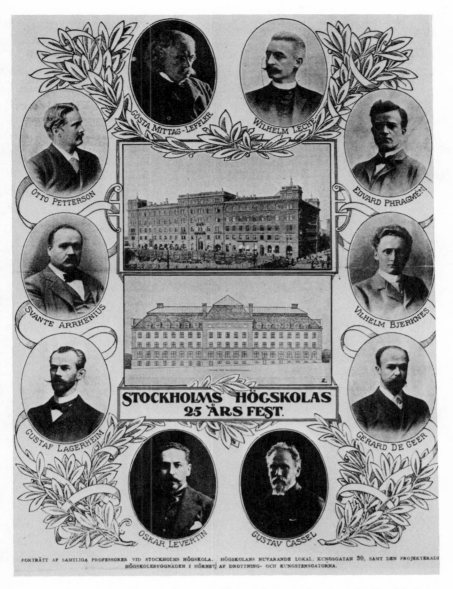

To commemorate its twenty-fifty anniversary in 1903, the Stockholm Högskola distributed this plaque with portraits of the entire professorial staff, ten in all. In the center is shown the building in central Stockholm where the Högskola rented offices and, below it, the projected Högskola building, which was inaugurated in 1909. (Courtesy of the Stockholm City Museum)

research interests of the staff and their more pronounced international orientation. As a new institution, the Högskola was also fortunate in having been able to recruit high-level scientific talent. Here, the actions of Mittag-Leffler had been particularly important since they contributed not only to the international standing of the mathematics department, but also to the strong position of mathematical physics at the Högskola. The latter was represented by a separate department headed by Bjerknes, whose work on hydrodynamics and electrodynamics, carried out in part in collaboration with his father C. A. Bjerknes, had attracted international attention.[35] In the mathematics department, Fredholm lectured on the partial differential equations that had important applications to problems in physics; in 1900, he published the general solution of the integral equation that established his reputation abroad.[36] Physical chemistry was another area where the Högskola was in tune with international developments. It received strong emphasis in the physics department, where Arrhenius held the chair from 1895 to 1904; as already mentioned, this made the department a magnet for foreign postgraduates. Physical chemistry also represented one of the many research interests of Pettersson, who headed the chemistry department and had been one of the earliest supporters of Arrhenius's theory of electrolytic dissociation in Sweden. Although by the 1890s Pettersson's primary research interest was hydrography, physical chemistry continued to be an important part of the work in his department. Uppsala, by contrast, lagged behind in the introduction of modern physical chemistry, with no chair until 1912.[37]

The building up of the Högskola was an arduous process that caused a series of conflicts culminating in the mid-1890s. Since these conflicts are thought to have influenced Nobel when he formulated his last wishes as well as to have affected the manner in which these wishes were carried out, they deserve brief mention here. The underlying cause of the disagreements at the Högskola were the conflicting views of two factions as to the basic aims of the institution. One faction, led by Mittag-Leffler, consistently supported *Lehrfreiheit*, that is, the view that the Högskola should be devoted to free learning and research at the highest level and not concern itself with exam or degree requirements. The other, led by Pettersson and including Arrhenius and Bjerknes, wanted to see the Högskola develop into a full-fledged university with the right to grant degrees in the same range of subjects as the state universities. The election of Pettersson as rector in 1893, followed by that of Arrhenius, who held this post from 1897 to 1901, were significant victories for this group. They were won, however, only after head-on confrontations with the old guard led by Mittag-Leffler. Pettersson's reelection as rector early in 1895 only occurred after a drawn-out fight during which the Högskola for a short while, partly as a result of Mittag-Leffler's machinations, found itself with *three* rectors, two of whom had been duly elected by the council of teachers at two successive meetings, the third having been appointed temporarily by the governing board.[38] This episode was followed only four months later, in May 1895, by a row over the appointment of Arrhenius to the newly created chair in physics. In an effort to block Arrhenius, Mittag-Leffler succeeded in having Lord Kelvin called in as one of the three experts judging Arrhenius's competence. Kelvin's opinion was predictably negative,[39] but was transmitted to Mittag-Leffler in a personal letter rather than, as called for by the rules of appointment (*sakkunnighetsinstitutionen*), in a formal opinion addressed to the Högskola. This made it possible for Pettersson, the new rector, to disallow it. Since Hasselberg, another of the three experts, had only expressed a tentative approbation of Arrhenius's qualifications, the appointment of the latter was put in jeopardy. That the board finally decided in Arrhenius's favor was probably due, in equal measure, to Pettersson's spirited defense of his friend and to the testimonials gathered together from a galaxy of German scientists, including L. Boltzmann, W. Hittorf, F. Kohlrausch, W. Ostwald, and E. Warburg.[40] As we shall see in the following chapter, this episode indirectly influenced the procedures for involving foreign scientists in the evaluation of candidates for the Nobel prizes.

In Stockholm's small intellectual community, where practically everybody knew someone in the press, these conflicts received considerable attention from the newspapers, which made much of "the trouble at the Högskola." The news often emanated from the contenders themselves, who did not shy away from press campaigns and pamphleteering if these could help gain support for their opinions. This unflattering publicity has been cited as the reason Nobel bequeathed no part of his fortune to the institution. Although it is not known how those in responsible positions at the Högskola came to believe that a *large* bequest was forthcoming, this indeed was the expectation, and the disappointment was keen when it was announced early in 1897 that the Högskola had been left out of Nobel's final will of 1895.[41] Recriminations followed, with both Pettersson and Arrhenius letting it be known that Nobel's dislike for Mittag-Leffler had brought about what Pettersson termed the "Nobel flop."[42] This is only of interest because it may have contributed to the myth that Nobel had planned to institute a prize in mathematics but refrained because of his antipathy for Mittag-Leffler or – in another version of the same story – because of their rivalry for the affections of a woman.[43] A more important rivalry, however, was the one between Mittag-Leffler and Arrhenius. Its impact on Nobel prize decisions will be discussed in Chapter 5.

The Karolinska Institute

During the period under discussion, the most significant expansion of staff and facilities at any scientific institution occurred at the Karolinska Institute. Since this was also where Nobel placed the awarding of the prize in physiology or medicine, its organization and status will be described briefly. The Karolinska had been created by royal decree in 1810 in response to the urgent need for trained surgeons in the army that was then involved in fighting the Napoleonic wars. This initial task was a short-lived one, however, and the institute soon started to accept students for general training in clinical medicine. In this capacity, it took over the functions of the Collegium Medicorum, founded in 1663, which had responsibility for the practical, especially surgical, training of doctors in the capital. Initially, the institute only had the right to grant the degree of master of surgery, required of doctors entering the army or the civil service, while the medical faculties at Uppsala and Lund kept the prerogative of granting such general degrees in medicine as *medicine kandidat* and *medicine licentiat*. Also, the medical education dispensed at the universities was largely abstract and science-oriented, whereas that of the institute was more practical and involved hospital training. In the course of the nineteenth century, the institute, despite strong opposition from the universities, introduced courses in most areas of medicine and successively won the right to grant the degrees of *medicine licentiat*

(1861) and *medicine kandidat* (1874) and to examine theses for the degree of *medicine doktor* (1886). Having acquired the status of a medical faculty, the institute experienced rapid growth in the number of professorships, which rose from eight in 1870 to twenty-four in 1914. Its research orientation also became more pronounced and was well known to Nobel who, being interested in experimental medicine, had been in contact with professors at the institute.[44]

Research orientations and "schools" in Uppsala and Stockholm

The above description of teaching and research in the departments that provided the bulk of the membership of the Nobel Committees for Physics and Chemistry gives some preliminary insight into the research orientations of the men who became responsible for the selection of Nobel prizewinners. The specialties of these individual committee members would have a great deal to do with the selection of laureates, but the more pervasive influence over prize selections stemmed from shared research orientations and philosophies of science. When these become the common denominator for members of a group it is possible to speak of a "school." This was so in physics, where representatives of the Uppsala school of experimental physics (Thalén, Ångström, Hasselberg, and Granqvist) dominated the Nobel committee during its first decade.

The Uppsala tradition in experimental physics developed under four successive chairholders who, together with their students, worked in a few closely related research areas. The *Leitmotif* of their work was the ambition to perform precise measurements and to determine natural constants in the area of radiation emission and absorption, primarily by means of spectroscopy. The founder of the school was A. J. Ångström (1814–1874), who, among other investigations considered classic in the field, undertook the detailed mapping out of the solar spectrum. Ångström was honored for this work, which formed the basis for the standard scale used in spectroscopic measurements, when the angstrom unit was named after him. Ångström's work was continued after his death by Thalén (1827–1905), who prepared the later work of his predecessor for posthumous publication. He subsequently took up the difficult problem of identifying the characteristic spectra of the

metals of the rare earth elements (scandium, thulium, and ytter-
bium) discovered by Cleve and L. F. Nilson. This line of investigation was continued by Hasselberg, Thalén's student, who, as
Academy physicist, was engaged for some fifteen years in revising
and refining spectroscopic determinations of a large number of
metals.[45] The third generation was represented by Knut Ångström
(1857–1910), son of A. J. Ångström, who specialized in solar radiation and improved considerably on existing measurements of
the solar constants by using instruments he designed himself, among
them the compensating pyrheliometer. Ångström's investigations
were ended by his untimely death, and it was left to his successor,
G. Granqvist (1866–1922) to carry these forward. In any event,
Granqvist became too preoccupied with teaching and administration to make a contribution in this area of research.[46]

Although the work of the three generations exhibits continuity,
there were also significant differences in their research styles.
Whereas the work of the older Ångström ranged over the whole
area of spectroscopy and concerned not only experiments, often
carried out with improvised means, but also important theoretical
formulations, that of the second and third generation was narrower
in scope and primarily concerned with exact measurements. In
this respect, the Uppsala physicists played their part in the international discipline of physics, where the emphasis placed on experimental techniques and precise measurement represented a shared
value as well as the means whereby common standards and a
common language in the reporting of research results were attained. At the turn of the century, experimental physics was most
highly developed in Germany, but it had also become the model
for disciplinary organization elsewhere.[47] Given the traditional emphasis on natural history and the long reign of idealist philosophy
in its universities, however, Sweden had lagged behind Germany
in efforts to modernize physics teaching and research. Perhaps it
was because the second and third generations of experimental physicists at Uppsala had to devote so much time to improving facilities, equipment, and training that they set their research
ambitions rather low, formulating problems primarily in terms of
technique and instrumentation and avoiding theoretical work.[48]

The notion of a "school" implies not only a convergence of
research interests and a shared philosophy of science but also a

collective identity that gives social cohesion to the school, thus enforcing the loyalties that make members act as a group in laying claims to positions of authority and power within a discipline.

The group identity of the Uppsala school was mainly derived from the research orientations the faculty had in common with experimental physicists in Germany and other countries. However, local conditions (e.g., the influence of key individuals and institutional rivalries) characteristic of a small and peripheral scientific community led to some selectivity in the emulation of general trends. Thus, the positivist and empiricist stance of experimental physics was exaggerated, making for an "experimenticist" philosophy of science that, according to Holton, "is best recognized by the unquestioned priority assigned to experiments and experimental data in the analysis of how scientists do their own work and how their work is incorporated into the public enterprise of science."[49] The experimenticist stance was the basis of the Uppsala school's definition and delineation of itself within Swedish physics. It was expressed, often in exaggerated form, in programmatic statements introducing research reports, in biographical memoirs (levnadsteckningar) of fellow members of the Academy, as well as in the low esteem in which colleagues who were not skilled experimenters were frequently held.[50]

It is noteworthy that these differences in scientific philosophy also prevailed within the field of spectral studies in Sweden. Thus, Uppsala spectroscopists saw no connection between their work and that of Janne Rydberg (1854–1919), professor of physics at the University of Lund, who organized spectral series data into the general mathematical formula that proved of great value in developing theories concerning atomic structure. One of Rydberg's biographers has cited his primarily mathematical approach and highly individualistic personality as reasons for his isolation from both Swedish spectroscopists and other physicists in Lund.[51] In fact, Rydberg enjoyed a higher reputation abroad than at home. The low esteem in which Hasselberg held the work of Rydberg and others who had launched into "theorizing" before accurate measurements of the spectra of all the elements had been obtained is apparent from his dismissal of the results of theoretical studies as being "so insignificant that their importance for the main question [i.e., the general laws producing different types of spectra] is put in doubt."[52]

Members of the Uppsala school were drawn closer together by the personal and intellectual rivalries that separated them from the scientists at the Högskola. It is possible that they would have taken a less restricted approach to physics had it not been for the fact that the scientists at the Högskola claimed to represent much of what was new in physics research in Sweden. Furthermore, many Uppsala physicists had a personal dislike for Arrhenius.

That Uppsala professors should have perceived the Högskola not just as an upstart institution but also as a threat to their university is understandable when viewed in the context of the rapid growth of the Högskola and the frequent assertions of its leaders that the capital was the proper place for a large, modern university oriented toward the sciences.[53] More important, the prevalent intellectual climate at the Högskola ran counter to the specialty and disciplinary loyalties evinced by the Uppsala physicists and may have produced a retrenchment on the part of the latter. There is ample evidence in the private correspondence of Arrhenius and Pettersson that they saw their role as being in the vanguard of science, particularly when it came to the breaking down of disciplinary boundaries through hypotheses linking theories developed in one discipline to problems encountered in another. There is no doubt that they considered this a more fruitful approach than the work of those who did not stray from well-defined research areas and methodologies.

It was unusual for general differences in scientific philosophy to be aired publicly, but when Pettersson spoke before the Swedish Society of Chemists shortly after the Nobel prize award to Arrhenius in 1903, he must have felt that the moment had come for a head-on attack on the tendencies that, in his opinion, had made the Uppsala physicists, in particular, oppose Arrhenius's work. These tendencies were twofold: one was what he termed (altering a German word for his purposes) *fack-filisteri*,[54] and the other resistance to the use of hypotheses in scientific research. He defined *fack-filisteri* as "the tendency to sort everything into certain predetermined categories and to neglect interjacent areas," and pronounced it "one of the more unbearable habits of the scientific world." The resistance to the use of hypotheses he regarded a consequence of "intellectual myopia," which, in its positive aspects, makes the scientist set the highest standards of verity for his work. There is a negative side, though, in that the striving for accuracy makes many scientists restrict their research to descriptions of what they see through their microscopes. Thus, they become incontestable authorities in very small areas of research. Should someone leave his domain of expertise, he is immediately attacked for lack of thoroughness (*grundlighet*). Thus "the researcher becomes like the Pope; he cannot make a move without losing his position of authority!"[55]

The conflicts between the Uppsala school and scientists at the Högskola must be viewed in the context of the multidisciplinary program in the physics and chemistry of the earth and atmosphere carried out in the 1890s by Högskola scientists (among them Arrhenius and Pettersson) and in discussions at the Stockholm Physics Society. Although its direct influence on Nobel prize deliberations was marginal, this program deserves brief mention since it illustrates the different orientations of physics in Stockholm and Uppsala.

The overall aim of the Stockholm program was to establish the unity of the physical sciences by using hypotheses and data to link geophysics, atmospheric science, astrophysics, and evolutionary cosmology. The Högskola scientists, who represented all of these fields, referred to this effort as "cosmical physics." As this term indicates, it was a physics of macroscopic scale, since the phenomena they sought to explain were such massive ones as magnetic storms, volcanic activity, the aurora borealis, glacial epochs, and eventually, in the case of Arrhenius, the development of galaxies and stars. Because these and similar phenomena are too distant in time and space to permit direct observation, their description and explanation was attempted through analogical reasoning, using models derived from mechanics to infer from the known to the unknown – or, rather to the "not yet known." This is best illustrated by Bjerknes's ambitious project in weather prediction in which he proposed to define the state of the atmosphere at any given time and then use appropriate hydrodynamic and thermodynamic laws to calculate future states. This method was to give birth to the Bergen school of meteorology founded by Bjerknes after the First World War.[56]

The most important stimulus to the construction of the many theories and models subsumed under "cosmical physics" came from the rapid accumulation of data not only on the physical and chemical composition of the earth and its atmosphere, but also as a result of the introduction of spectroscopic methods into astro-physics on the atmospheres of other planets. Thus, the work of the Högskola scientists had an empirical basis – data that they themselves and others had obtained in the different specialties con-cerned. But much of their work (particularly when discussed in the Physics Society) also showed, as Hiebert has pointed out, that "physicists around 1900 were caught up in a spirit of scientific speculation that would have been considered to be quite reckless ten years earlier."[57] Needless to say, for the Uppsala physicists, this work was much too speculative. In particular, Arrhenius's studies of atmospheric electricity were severely criticized by Has-selberg in the latter's expert opinion (1895) on Arrhenius's com-petence to hold the chair in physics at the Högskola. This was also the beginning of the enmity that characterized their relation-ship. As Arrhenius turned his attention to cosmogony and de-veloped the theories reported in the book *Worlds in the Making*, many of his friends would probably have agreed that the frankly speculative side of his work had taken the upper hand.[58]

Although the orientations of the Uppsala physicists and the scientists at the Högskola were so different, they were both to some extent responses to the lack of adequate resources and fa-cilities for research. Being state-financed, and hence able to count on a stable and even growing subsidy, the Uppsala physicists could bide their time, concentrating on their areas of strength in exper-imental physics. The fact that the Högskola during much of the 1890s was unable, for financial reasons, to equip itself with the apparatus required for specialist work in experimental physics or chemistry probably contributed to the far-flung research interests of its teachers.[59] The scientific daring of Arrhenius and Pettersson, in particular, was matched by a similar innovativeness when it came to institutional arrangements that could benefit Swedish sci-ence and, indirectly, the Högskola. Here, the ideas of both men would influence the Nobel institution in its earliest, formative stages, when its goals and purposes were being defined and the guidelines for its independent functioning drawn up.

3

Implementing the will of Alfred Nobel, 1896–1900

Alfred Nobel died in his villa in San Remo on December 10, 1896, and his will (dated November 27, 1895), which had been deposited in the Enskilda Bank in Stockholm, was opened five days later. It is no exaggeration to say that the contents of the will came as a complete surprise to those most directly affected by its major provisions. The most unpleasant surprise was of course sprung on Nobel's closest relatives, principally his two nephews Hjalmar and Emmanuel Nobel, who learned that the bequests made to them and other members of the family represented only a limited portion of the estate: 1 million crowns out of assets estimated at over 30 million crowns. They also learned that no member of the family had been named executor to the estate. Despite the disappointment that this news must have caused him, Emmanuel Nobel, who represented the Russian branch of the Nobel family, felt that he should not contest his uncle's expressed wishes; the support he gave them was later instrumental in putting the will into practice.

The provisions in the will also brought surprise and dismay to Ragnar Sohlman (1870–1948), the young Swedish engineer employed as Nobel's personal assistant, who learned in San Remo that he together with Rudolf Lilljeqvist (1855–1930) had been appointed executors of the estate. Sohlman was only too well aware of the formal defects of the will, which named as the principal legatee a fund that did not exist and therefore had to be organized. He also anticipated the complex array of legal, political, and organizational issues that would have to be confronted once the will had been declared valid. In his memoirs, Sohlman observed, not without irony, that there was only one point on which the will was perfectly explicit, "and that was the appointment of Lilljeqvist and myself as executors." The major problems they faced could be summarized as follows:

1. Legal formalities and contentious matters involving, first, the problem of jurisdiction over the will, a very thorny question since after having left Sweden as a child, Nobel had never been a legal resident of any country and, second, the possible lawsuits that might be brought contesting the validity of the will or, in other ways, making claims against the estate.
2. Economic transactions connected with the liquidation of the property and the re-investment of all proceeds in the "safe securities," that were to constitute the capital of the fund referred to in the will.
3. The organization of an administrative body for the permanent management of the fund and the formulation of rules for the annual distribution of the prizes.[1]

The most important steps in setting up the Nobel institution were those taken with respect to the third set of problems, and these will be the focus of this chapter. The tangible end results were the statutes, promulgated in 1900, that provided the legal and organizational structures of the institution: the Nobel Foundation as the body managing the fund and the principles and procedures guiding the prize-awarding institutions. The lengthy negotiations over these will be described in some detail, for they involved defining, formally as well as informally, the roles of the different participants in the process of selecting Nobel prizewinners: the individuals invited to nominate candidates for the prizes, the Nobel committees, the proposed Nobel Institute, and the bodies making the final decisions. Representatives of the Academy of Sciences and the Karolinska Institute took the lead in advancing proposals and counterproposals; it is therefore not possible to dissociate discussion of the science prizes proper from that of the prize in medicine or physiology.

Serious negotiations about the statutes could not start, however, until all contentious matters involving the will had been resolved. The two years of battle over the will (1897–1898) deserve attention since the arguments engendered the view that the prizes were a public enterprise of importance to Swedish society and that to implement Nobel's wishes private interests and concerns had to be relinquished. This view was adopted by the executors and their attorney in defending the will against both the obstructions of Nobel's relatives, and the vacillations of the future prize awarders. Their success laid the groundwork for the creation of the Nobel institution as an entity guided by principles common to all of its constituent bodies and hence to some extent shielded from the special interests of each of the prize-awarding institutions.

STRATEGIC MOVES IN THE BATTLE OVER THE WILL

In late December 1896, a first meeting was held in Stockholm between the three men, none of whom had met before, who had primary responsibility for implementing Nobel's last wishes. They were Sohlman and Lilljeqvist, the two executors of the estate, and Carl Lindhagen (1860–1946), whom they had engaged as their attorney. The specific decisions they made at this time added up to an overall strategy for the battle over the will that was certain to ensue, given the intention of some of Nobel's relatives to have it declared void.

Sohlman and Lilljeqvist were Swedish engineers, both of whom had worked abroad. Sohlman became Nobel's personal assistant in 1893 and rapidly gained the trust of his employer. Lilljeqvist had worked as an engineer in England and France but returned to Sweden to launch the Electrochemical Company, for which Nobel had put up one-third of the capital when it was founded in 1895. Being fully occupied with the company, which was located far from Stockholm, Lilljeqvist's role as an executor was limited to offering advice and counsel. It was thus the twenty-seven-year-old Sohlman who had to give overall direction to the complicated enterprise of settling Nobel's estate and carrying out the many tasks this involved. Being familiar with Nobel's business organization and personal affairs, he became most directly involved with the first two matters discussed in the preceding summary. In 1897, he spent most of his time on the Continent arranging for the liquidation of the securities held by Nobel and for the sale of the properties in Paris and San Remo. He was also battling against the attempts of Nobel's Swedish relatives to contest the validity of the will in French courts.[2] During this time, Lindhagen acted as the chief representative of the executors and the estate in Sweden.

Ragnar Sohlman (1870–1948), one of the two executors to Nobel's estate (*c.* 1895). (Courtesy of the Nobel Foundation)

Carl Lindhagen (1860–1946), attorney to Nobel's estate (*c.* 1900). (Courtesy of the Royal Library)

The most important early decision taken by the executors was probably to engage Lindhagen, a judge referee at the Court of Appeals, as the attorney for the estate. Having risen rapidly in the court system, Lindhagen was then entering political life and became a member of Parliament in 1897. Officially, he represented the Liberal Party, but his interests in social and political reform were too wide-ranging to be fitted into the party mold. These included women's political and legal rights, land reform, the socioeconomic development of northern Sweden, and the property rights of the ethnic minority of Laplanders. Lindhagen brought to the work of settling the estate the same energy and innovative spirit that he invested in social and political reform. He acted at the same time as lawyer, negotiator, and architect of the detailed rules and procedures for the awarding of the prizes. He exerted a strong influence over the science prizes and was particularly well placed to do so since his family connections and friendships gave him privileged access to both the more traditional and the more dynamic parts of Swedish science. His uncle, Georg Lindhagen (1819–1906), an astronomer and professor at the Observatory of the Academy of Sciences, had served as secretary of the Academy since 1866.

More important, Lindhagen's father, Albert (1823–1887), was one of the founders of the Stockholm Högskola and acted as its secretary until 1884 when he was succeeded by Carl, who only the year before had earned his law degree from the University of Uppsala. Carl Lindhagen remained in this position until 1894. Both then and later he was active in securing and expanding the financial base of the Högskola and in planning its future growth. To further these ends, he worked closely with Pettersson and Arrhenius, both of whom were close friends from his student days in Uppsala.[3]

Turning now to the strategy initiated by the executors and their attorney early in 1897, one of their first decisions was to try to have Nobel declared a legal resident of Sweden, thus giving Swedish courts formal jurisdiction over the estate. This was not just a legal maneuver to avert the threat of the will being declared invalid by a French court. It formed part of a series of actions designed to bring forth not only legal but also political support for the will – in short, to make the Nobel prizes a matter of Swedish national concern. To implement this strategy it was first necessary to secure the cooperation of the institutions named in the will as recipients of the bequest (the Royal Academy of Sciences, the Karolinska Institute, the Swedish Academy, and the Norwegian Storting). Then, Lindhagen proposed in a letter to Sohlman in early January 1897, that the government be involved in drawing up statutes for the fund and perhaps even persuaded to take on its management.[4]

The third component of the strategy that was taking shape in early 1897 involved the press. Informing the public of Nobel's last wishes was an urgent matter that had to be handled carefully so that the major newspapers would support the will, thus making the members of the prize-awarding institutions aware of their responsibilities to the public and the nation. Sohlman and Lindhagen could count on making their views known through one major paper, the liberal *Aftonbladet,* of which Sohlman's brother Harald was editor-in-chief. Being active in the Liberal Party, Lindhagen would also have access to other newspapers with this political orientation. There was every reason to believe, however, that the Swedish relatives of Nobel would mount·their own campaign, using other newspapers.

The first newspaper to publish the will was *Nya Dagligt Allehanda* in Stockholm, on January 2, 1897. In the next few days, the news was carried by the bulk of the Swedish press and also spread abroad. A count of the number of press clippings forwarded to the executors by the French agency that had been retained to monitor the world press reveals that between January and May 1897 nearly one hundred newspapers in Europe, North America, and Asia printed the news about Nobel's bequest. In Sweden, the press was almost unanimous in its praise for Nobel's intentions, even if some doubted their practicability, and in no little awe of the enormous sums that would be given away in annual prizes. In *Nya Dagligt Allehanda,* for instance, Nobel's gift was declared "probably the most magnificent one of its kind that a private person has ever had both the desire and the ability to make." The sum to be awarded in annual prizes, estimated at a total of about 1 million crowns ($270,000), was referred to as a "fortune" – and indeed it was in the monetary terms of the time. With each prize assumed to be around 200,000 crowns ($54,000), it represented about thirty times the annual salary of a university professor and two hundred times that of a skilled worker in the construction industry.[5]

More important to the defense of the will, however, was the fact that many newspapers regarded the Swedish *nation* as the chief beneficiary of Nobel's bequest, since this belief, it could be hoped, would give the future prize awarders a sense of carrying out a historic mission. Many newspaper accounts of the bequest had nationalistic overtones. In one paper (*Svenska Dagbladet*) this was taken to extreme lengths. Comparing the competitition for the prizes to the Olympic Games, Sweden became the "Olympia" where each year men of knowledge from the whole world would present their most brilliant achievements: "Swedish scientific societies could not have been invested with a more glorious task, and one involving a larger measure of responsibility, than to serve as the Aeropagus of these worldwide prize competitions involving the foremost intellectual forces of humanity."[6] Only one paper, the conservative *Göteborgs Aftonblad*, struck a discordant note in this chorus of praise. In a series of articles, probably inspired by the heirs contesting the will, Nobel's intentions were denigrated and described as an expression of "superficial cosmopolitanism." The paper suggested that Nobel's relatives and representatives of the prize-awarding institutions work out a compromise that would satisfy both the rights of the heirs to a more equitable share of the estate and the interests of the institutions named in the will.[7]

The most important task for Lindhagen, once the press campaign had been launched, was to secure the cooperation of the institutions designated to award the prizes. It was necessary for the executors, he wrote to Sohlman, to remain in control of events. This required that the institutions be officially informed and presented with a course of action. Such an official initiative would have to await Sohlman's return, however, and in the meantime Lindhagen could do no more than maintain informal contacts with influential members of the institutions. His reports on these contacts in letters to Sohlman show that the munificence of Nobel's bequest had made a strong impression on members of the scientific community. Professor A. Key, rector of the Karolinska, whom Lindhagen had met in the street, held forth about "the splendid chancellery and library that each prize committee would need." Professor Nordenskiöld, the Arctic explorer and an influential member of the Academy of Sciences, had expressed the opinion that "each member of the prize committees should receive a professor's salary."[8]

The more disturbing news that he reported to Sohlman was that Key had expressed support for the proposal that an agreement be reached with Nobel's heirs giving them possession of the whole fortune (estimated at 30 million crowns), if in return they put, say, 25 million crowns at the disposal of the institutions. Although the origins of this proposal are not discussed in the letter or elsewhere, a document found in Lindhagen's papers gives a clear indication that plans for just such an agreement were being made by those close to the prize-awarding institutions. This document indicates that the annual proceeds of such a fund would likely represent a sum (700,000 crowns) almost as large as the budget of Uppsala University in 1895. This sum would be used to create laboratories and other research facilities in the areas of science, medicine, and the humanities. These facilities would be known collectively as the Nobel Institute and would be open to scientists and scholars regardless of nationality. The proposed institute would also proffer grants and award annual prizes on a more modest scale than envisaged by Nobel.[9]

To bolster the position of the executors, Lindhagen sought the opinion of the Swedish attorney general on the prospects for government action in defense of the will. The attorney general took a positive view but declared that he wanted to consult his colleagues. He asked Lindhagen to submit a confidential memorandum outlining the course of action that the executors proposed to follow.[10]

THE SKETCHING OUT OF THE STATUTES

With all the parties to the dispute mobilized, it now became urgent for Lindhagen and the executors to reach agreement on the basic principles to be included in the statutes for the fund and on the procedures that should be followed in drawing up the statutes. These procedures were spelled out in the document entitled *Plan* that Lindhagen drafted and that also formed the basis of his memorandum to the attorney general. Lindhagen proposed that when the institutions were to be asked to accept responsibility for awarding the prizes, they should at the same time be informed of the executors' intention to request that the statutes be promulgated by the government. The statutes would then be drawn up by the executors in collaboration with the appointed representatives of the institutions. Before the statutes were to be submitted to the government, the approval of Nobel's heirs would also be sought – "but not bought."

The main part of the *Plan* contained the more important provisions to be included in the statutes. With some exceptions, the principles set forth were the same as those contained in two documents laying the groundwork for the statutes drafted by Sohlman.[11] Both Lindhagen's *Plan* and Sohlman's second, more elaborate document, stated as a first principle that the wording of the will should be followed literally. It would be necessary, though, wrote Sohlman, to include provisions in the statutes concerning eventualities not foreseen in the will and to clarify some of the will's language. The following list outlines the proposals for major modifications and clarifications in the order in which they appeared in the documents referred to above.

1. It was deemed necessary to clarify the stipulation found in the will that the prizes be awarded to "those who, *during the preceding year*, shall have conferred the greatest benefit on mankind." Lindhagen's *Plan* stated that in order to be considered for the prize, a discovery would not have had to be made during the preceding year but that it was sufficient if its importance had been established in that year. This principle was subsequently inserted into the statutes (para. 2). (The texts of the statutes of the Nobel Foundation and of the special regulations for the prizes awarded by the Royal Academy of Sciences are reproduced in Appendix B.)

2. Both Lindhagen and Sohlman agreed that in certain circumstances, a prize could be divided into two or more awards. Sohlman specified that this might occur "if two or more works of equal scientific or technical merit are presented to the committee for prize adjudication."

3. Both men addressed the question of how to proceed in case no discovery deemed worthy of the prize had been produced or recognized during the year. This eventuality, which Nobel had neglected to consider in his will, was to play a prominent role in subsequent discussions of the statutes since it opened up the possibility that the proceeds of the fund, if not always used for prizes, could benefit the awarding institutions. This possibility was introduced in Lindhagen's *Plan*, which set forth the principle that "prize money which cannot be distributed should be used in another manner" and that such a use should be in conformity with the purposes that the prizes were intended to serve.

4. The memoranda dealt with the procedures for selecting laureates. Sohlman proposed that the prize-awarding institutions elect juries (*nämnder*) each comprising three members – Swedish or foreign – who would judge the works being considered for the prize. No reference was made, however, to the manner in which these works would be brought to the attention of the juries.

In their first sketching out of the statutes, Lindhagen and Sohlman do not seem to have used the rules of any existing prize funds as models. "It is always better," wrote Lindhagen to Sohlman, "to draw up statutes independently of existing ones."[12] The groundwork for the statutes was also laid, probably intentionally, without the participation of the institutions that would be responsible for awarding the prizes. Once brought into the drafting process, though, the future prize awarders could be expected to bring important modifications to the ideas expressed in these first memoranda.

On March 24, 1897, the first step in the negotiations with the prize-awarding institutions was taken by the executors. In a letter addressed to the Academy of Sciences and sent to the other institutions as well, the executors asked to be informed whether the Academy would be willing to award the prizes in physics and chemistry. The Academy was also invited to name representatives who, together with the executors and one representative from each of the other institutions, would draw up the detailed provisions for awarding the prizes. To prepare an answer to the letter the Academy appointed a committee of five members that, among other matters, was to present the Academy with a proposal specifying the conditions under which it would agree to award the prizes. Among the members appointed to the committee were D. G. Lindhagen, B. Hasselberg, and L. F. Nilson.[13]

To provide guidelines for those who were formulating the reply of the future prize awarders to the executors' letter, Lindhagen drafted an "Explanatory Addendum to the Will of Alfred Nobel" in which he elaborated on his earlier *Plan* as well as on Sohlman's two documents.[14] That Lindhagen's memorandum influenced the deliberations of the Academy is shown by the fact that its main points were incorporated into the committee report submitted to the Academy in early May. This report recommended that the Academy honor the executors' request to appoint delegates to draw up rules for the management of the fund and the awarding of the prizes.[15] Lindhagen's explanatory addendum also figured prominently in the letter of May 29 in which the Karolinska Institute accepted the task of awarding the prize in medicine or physiology.[16] In both cases, however, the acceptance was provisional (the Academy committee's recommendation had of course to be ratified by the parent body), since it was subject to the satisfaction of certain conditions, specified in the replies, in the detailed provisions for the fund and the prizes. These conditions, which dealt with the compensation that the prize-awarding institutions should receive from the fund, were basically of two kinds and represented the initial bargaining positions of these institutions.

First, the institutions demanded that after the annual income from the fund had been divided into the five lots that were to constitute annual prizes, a certain share of each lot be set aside to cover expenses in connection with the selection of prizewinners. Lindhagen's memorandum had proposed that this share amount to 5 percent of each lot. From the beginning, however, the demands of both the Academy and the Karolinska ran considerably higher. Their ultimate success in pressing these is measured by the provision in the statutes (para. 13) that 25 percent of each lot be set aside to meet expenses and that the money not used for this purpose could be put in reserve by the prize awarders for their future needs. In the first year of the awards, this share amounted to about 45,000 crowns for each lot, only part of which was used, mainly for the honoraria paid to members of the Nobel committees. In the Academy of Sciences, these honoraria were set at 2,000 crowns annually or about one-third of the "professor's salary" that Key had felt was due to members of the prize juries.[17]

The second demand concerned the use of the prizes that had not been awarded due to a lack of worthy discoveries or inventions. Both the Academy and the Karolinska seized the opportunity of making suggestions for possible alternative uses of the prize money. That the institutions would be compensated in more ways than just having their expenses covered was an idea that had already been raised in connection with a possible settlement with Nobel's Swedish relatives, when it had been suggested that part of Nobel's donation be used to create the Nobel Institute mentioned previously. It was now introduced formally in the replies of both institutions.

The report of the Academy committee recommended that the savings accumulated from reserved prizes be used to establish "a well-equipped Nobel Institute for physics and chemistry [at the Academy] to which foreigners, too, should have access in order to carry out scientific investigations." This idea was echoed in the letter from the Karolinska, which proposed the creation of "scientific medical institutes, open to scientists of all countries, receiving the material means to pursue their investigations . . . as well as economic support enabling them to devote themselves fully to these." There will be occasion to return to the proposed Nobel Institutes, since they were to figure prominently in the settlement not, as one might have expected, with Nobel's Swedish relatives but with the Russian branch of the family headed by Emmanuel Nobel.

By late May 1897, the strategy devised by the executors and their attorney seemed to be paying off. The Norwegian Storting had been the first to reply to the executors' letter of March 24 and had accepted the mandate of awarding the peace prize. This was followed by positive replies from the Swedish Academy (May 28) and the Karolinska (May 29).[18] The recommendations of the committee appointed by the Academy of Sciences indicated that a favorable reply would soon be at hand from that institution, too.

The most important breakthrough at this time, however, resulted from Lindhagen's efforts to involve the government in the procedures to have the will declared legally valid. Following Lindhagen's initial contacts with the attorney general, the executors had submitted a copy of the will to the government "for information . . . [and] in connection with whatever measures the government might consider appropriate." The government's decision on this matter, taken on May 21, was based on the legal opinion of the attorney general that although no direct interests of the Crown were involved, as the donation had been made by a Swedish subject for public purposes, it was appropriate that the government assist in putting the testator's wishes into effect. The decision was significant in that it made the government a party to any legal measures that would be taken in probating the will. The prize-awarding institutions were also instructed to join the executors in such actions, in which case they would be entitled to assistance from the attorney general. This decision was considered a major setback for Nobel's relatives, since in bringing suit against the will they would be confronting not only the executors but also the state.[19]

In view of these successes, the June 9 decision by the Academy of Sciences to delay any action concerning its role as prize-awarding institution until the will had been probated was a keen disappointment to the executors and their attorney. The Academy not only rejected the committee's recommendation that it accept the mandate on the conditions outlined above, but it also refused the executors' request for the appointment of delegates to confer with them and representatives of the other institutions concerning amplifications of the will. The Academy's refusal was particularly vexing since it came about as a result of the persuasive powers of one member, H. Forssell, an economic historian who managed to convince his fellow academicians that any discussion of clarification of the will at this stage would only encourage the relatives to start a court action.[20]

By refusing to discuss its role as a prize-awarding institution, the Academy of Sciences effectively blocked the probate of the will. In his memoirs, Sohlman makes the following grim assessment of the situation as it appeared at that time: "In order to get the will proved it was necessary, of course, that the prize-adjudicating institutions accept in advance their respective assignments, as otherwise the will would be null and void in one respect or another. And when one of the prospective prize juries refused even to appoint delegates to discuss the terms on which it would accept such a responsibility, no binding negotiations on the subject were possible. No progress could therefore be made in the matter." The stalemate caused the executors to make initial moves toward a negotiated settlement with the relatives. During the remainder of 1897, there were no further official discussions between the executors and the institutions.[21]

THE UNOFFICIAL NEGOTIATIONS OVER THE STATUTES AND THE SETTLEMENT WITH THE NOBEL RELATIVES

In an effort to break the deadlock caused by the decision of the Academy of Sciences, the executors and their lawyer decided in December 1897 to invite delegates from the Karolinska Institute and the Swedish Academy (of literature) for an informal conference about the provisions to be included in the statutes of the Nobel Foundation. The invitation was accepted, and the first meeting, held on January 11, 1898, was attended by K. A. H. Mörner, the new rector of the Karolinska; C. D. af Wirsén, the permanent secretary of the Swedish Academy; and, in an unofficial capacity, L. F. Nilson, who had chaired the Academy committee that had recommended conditional acceptance of the executors' proposal and who himself strongly supported it. At the first meeting, af

Wirsén insisted that, in view of the absence of delegates officially appointed by the Academy of Sciences, the deliberations must be of a strictly private nature. This made it possible for Lindhagen to invite Pettersson as a second unofficial representative of the Academy (Pettersson subsequently attended the four meetings held in January and February 1898). At least two of these meetings were also attended by Arrhenius, who was not a member of the Academy and whose name, for that reason, does not figure in the minutes.[22] The refusal of the Academy to enter into discussion with the executors therefore had the unanticipated consequence that it was being represented at such discussions by two members whose views went counter to its June 1897 decision. The unofficial delegates who, together with Lindhagen, had all been active in building up the Stockholm Högskola, were in fact primarily interested in the creation of a Nobel Institute for scientific research with strong links to the Högskola.

As a basis for the discussions at the meetings, Lindhagen and Sohlman had prepared a memorandum in which they proposed amplifications to the will closely following those elaborated in the documents of early 1897. In his introduction to that memorandum, Sohlman issued a stern warning to those who thought anything would be gained through a separate agreement with Nobel's relatives under which a part of the donation would be used for purposes other than those stipulated by Nobel. "It is hard to understand," he wrote, "that anything good could come out of such a disloyal action which lacks all legal foundation and which will meet with the strongest resistance from the executors."[23] That Sohlman found it necessary to issue such a warning shows that the executors were still very much preoccupied with the possibility that one or more of the institutions might make separate agreements with Nobel's relatives.

The main significance of the meeting did not lie in discussions about Sohlman's and Lindhagen's memorandum, however, but arose from two other events that, in the recollections of Pettersson, were intimately linked. These were, first, the elaborate proposal for the establishment of a large Nobel Institute advanced by Sohlman, Nilson, and Pettersson in the meeting of February 1; and second, the declaration by Emmanuel Nobel on February 11 that he wished to respect his late uncle's wishes as expressed in the will and did not intend to dispute its terms.[24]

In his memoirs, Sohlman gives a detailed account of how Emmanuel Nobel, as the representative of the Russian branch of the Nobel family, came to dissociate himself from the Swedish Nobel relatives who only ten days before his declaration had taken steps to start legal action to contest the will in a Swedish court. He cites, among other things, the close relations that Emmanuel had maintained with his uncle and the respect he felt for the ideals that had inspired Nobel in drawing up his will. In a series of meetings, Sohlman had been able to ascertain that Emmanuel's financial interests in the estate were of a limited nature. He wished to acquire for himself and his family the controlling interest in the Nobel Brothers Naphtha Company in Baku, which Alfred Nobel had retained, although from 1888 the management of the concern had been in Emmanuel's hands. The preliminary agreement that the executors reached with Emmanuel, allowing him to buy Nobel's shares in the company at a price advantageous to him, certainly played a major role in his decision not to contest the will.[25]

That the idea of a Nobel Institute also played a role in Emmanuel's decision is brought out in the letter that Pettersson wrote to Lindhagen in 1921, upon reading Sohlman's account of the creation of the Nobel Foundation published in the volume honoring Lindhagen on his sixtieth birthday.[26] Although Pettersson may have exaggerated the importance of an idea for which he had fought vigorously and lost, this hitherto unknown letter deserves to be quoted at some length. He writes:

In the creation of the Nobel Foundation, there was a critical point – a meeting between the executors and Emmanuel Nobel to which you also had invited me and Nilson, who was then member of the board of the Stockholm Högskola, in an advisory capacity....

The deliberations continued after a dinner which Nobel gave at the Grand Hotel and were followed by another meeting in which, at my suggestion and Nilson's, Arrhenius also participated. It was the first meeting that was decisive, however, and also the one that I remember best, since I, together with Nilson, was asked to summarize the proposal we had made concerning the creation of a *big* Nobel Institute with laboratories for research in physics, chemistry and physiology and a common library, all forming a *single unit*.... When I wrote my proposal, I envisaged the Stockholm Högskola as the closest neighbor to the big research institute up on the hill [Observatory Hill] and in close collaboration with it.... I think Emmanuel Nobel was fascinated by the idea and that his magnanimous renunciation of his own family interests depended, in some part, on the vision of a big Nobel Institute in Stockholm.[27]

"This was not to happen," Pettersson concludes this part of the letter, and he goes on to discuss the reasons, which will be examined below, why the idea of a Nobel Institute common to all the Swedish prize-awarding institutions had to be abandoned. It is easy to understand why representatives of the Högskola pressed hard for the creation of a large research institute in Stockholm since this fitted their long-term plans for a science-based university in the capital. In the short term, a Nobel Institute close to the Högskola would resolve some of the pressing problems of finding laboratory space for its teachers and students. Moreover, if some of the professors at the Högskola were also given positions at the institute, this would alleviate the permanently strained budget of the former institution. That Arrhenius had a personal interest in the Nobel Institute is indicated by a passage in a letter he wrote to Ostwald late in 1898. Complaining about how much of his time was consumed by his duties as rector of the Högskola, he stated, "Perhaps my situation will improve, however, when a Nobel Institute is created and I can leave the Högskola for the Institute," and added, "maybe the wonderful golden era when I only had to think about scientific tasks will come back."[28]

In the formal proposal for a Nobel Institute that Nilson and Pettersson made at the meeting of February 1, there was no mention, for obvious reasons, of any links with the Högskola. The novelty of the proposal lay in the dual rationale it presented for the proposed institute: on the one hand, its close links to the selection of prizewinners in science and medicine; on the other, its role in internationalizing Swedish science. With respect to the former, it was argued that due to the large sum of money to be awarded in annual prizes, the selection of prizewinners would require "detailed scientific investigations to determine the factual basis of a discovery and its significance." To do this, it would be desirable to have well-equipped laboratories where the work could be conducted by renowned scientists. That the proposed institute would, indeed, be *big* (to recall Pettersson's term) compared with contemporary Swedish scientific laboratories is indicated by the size of its staff. It was proposed that for each of the three departments (physics, chemistry, and medicine–physiology) there would be two or three directors and two assistants – that is, twelve to fifteen scientists in all.

The second argument advanced the notion that in order for Swedish science to acquit itself honorably of the task of awarding the prizes, it was necessary to arrive at a "critical mass," to use a modern term, of personnel and facilities. The main way to achieve this, according to the proposal, was through the establishment of an institute with a strong international orientation. The staff of the institute would be made up of foreign as well as Swedish scientists, and its long-term research program would be supervised by an advisory committee consisting of recent Nobel prizewinners. The authors of the proposal were aware of the fact that their ideas were novel and, in many respects, controversial. "It should not be surprising," they wrote, "that one seeks new forms for evaluating and rewarding scientific work when the unexpected event occurs that the largest scientific prize the world has ever known is entrusted to the academies in Stockholm."[29]

The document proposing the Nobel Institute contained another innovation: the idea that certain categories of scientists should have the right to nominate candidates for the prizes. In their earlier memoranda, Sohlman and Lindhagen had placed nominations as well as evaluations of candidates in the hands of juries composed of Swedish and foreign scientists and scholars. The idea of soliciting nominations from outside was forwarded by Arrhenius at the meeting of January 26, 1898. His point of departure was the principle, affirmed in the earlier memoranda, that the selection of prizewinners should not be based on applications. He suggested as an alternative arrangement that certain categories of scientists be given the right to nominate candidates. In the case of prizes awarded by the Academy of Sciences, this right would be vested in Swedish and foreign members of the Academy and could also be extended to the personnel of the science faculties at the universities in Uppsala, Lund, and Stockholm. In the formal proposal for the creation of a Nobel Institute, former Nobel prizewinners were added to the categories of nominators.[30]

The declaration of Emmanuel Nobel represented a breakthrough in the negotiations, since it meant that the share of the estate (eight-twentieths) to which the Russian branch of the family would have been entitled had the will been declared void could be used for the purposes intended by Alfred Nobel. The executors seized this opportunity to renew their request to the Academy of Sciences to name delegates for discussions concerning the provisions to be included in the statutes of the foundation. This time the Academy decided in favor of the request and named as its representatives L. F. Nilson and B. Hasselberg.[31]

The executors also continued discussions with the Swedish relatives, trying to negotiate a financial settlement that would end their opposition to the will. Ultimately, a series of agreements was reached in late May and early June of 1898 according to which the relatives dropped their suits in return for payments of about 1.5 million crowns over and above the legacies made to the family in the will. These agreements also stipulated that the statutes common to all prize-awarding bodies be drawn up in consultation with representatives of the Nobel family. When, late in 1898, the settlement had been accepted by all the prize-awarding institutions as well as by the Swedish government, the way had been cleared to begin official negotiations over the statutes.[32]

THE ELABORATION OF THE STATUTES AND SPECIAL REGULATIONS, 1899–1900

The group that met for the official negotiations over the statutes of the Nobel Foundation was very different in size and composition from that which carried out the unofficial discussions. The Large Nobel Committee, as the group came to be known, had thirteen members. Among these were Nilson and Hasselberg, representing the Royal Academy of Sciences; K. A. H. Mörner, the new rector of the Karolinska Institute; and af Wirsén, the permanent secretary of the Swedish Academy. Members of the Swedish and Russian branches of the Nobel family were present at the initial meetings but were subsequently represented by A. Posse, former Swedish prime minister, and H. Billing, a member of the Swedish Supreme Court. The only party not represented was the Norwegian Storting, which had decided that it would negotiate directly with the executors and their attorney.

After the preliminary meetings, at which he had been named secretary to the committee, Lindhagen began work on drafting the statutes. Judging from his letters to Sohlman, he seems to have done so without involving the executors since, in his own words, his draft would "constitute no more than a basis for discussion."[33]

The executors and the other committee members, however, considered the Lindhagen draft a much more finished document. If, as seems likely, Lindhagen's intention had been to give the executors and himself the upper hand in the negotiations, he achieved this. From the start, all discussion was carried out on the basis of his draft and all counterproposals took the form of a redrafting of its paragraphs.

The general principles contained in the statutes

Lindhagen's draft contained four main sections, on the purpose of the Nobel Foundation, the awarding of the prizes, the Nobel Institute, and the management of the fund. Although the constituent elements were brought together under the common heading of the Nobel Foundation, there was, in fact, a separation of power. On the one hand, there were the prize-awarding institutions, which would be involved mainly with the selection of prizewinners, and on the other, the board of the foundation, which would oversee the financial management of the fund. The different decision-making bodies were bound by very detailed provisions and regulations concerning all matters involving the fund and the prizes.

Lindhagen's draft was clearly the work of a lawyer or legislator who, when confronted with a matter requiring adjudication, first sets forth certain general rules and principles and then provides for machinery whereby these principles will be given concrete form through a series of acts and decisions. The general principles in Lindhagen's draft concerned the definition and demarcation of the various fields in which prizes were to be awarded, the ranking of works proposed for the prizes according to their importance, the dividing of a prize among no more than three persons, and the conditions under which a prize could be reserved.[34] The elements of the system most directly concerned with the prizes in science and medicine were the *nominators*, drawn from the personnel of Swedish and foreign faculties of science and medicine; the *committees*, one for each prize, whose members would be elected by the prize-awarding institutions and who would give their opinion on the nominations; the *Nobel Institute*, which, would assist in the evaluation of prizeworthy works; and the *prize-awarding institutions*, which would make the final decisions concerning prizewinners.[35]

That the negotiations over the statutes would be difficult became apparent at the first meeting of the Large Nobel Committee when the representatives of the Karolinska Institute (K. A. H. Mörner and R. Tigerstedt) presented a substantially modified version of Lindhagen's draft statutes as a counterproposal. The preamble to their version shows that they strongly disagreed with the notion that the different prize awarders be bound by common rules. "We do not think," it stated, "that the statutes should be drawn up in such a manner as to restrict the autonomy of the different prize awarders, at present or in the future, and to give one authority over the other."[36]

The Karolinska delegation argued that since the attribution of the prizes represented a new and complex area of activity, the statutes should contain as few detailed provisions as possible concerning the procedure that each prize-awarding institution would follow when selecting the prizewinners. These matters should be dealt with, instead, in the special regulations drawn up by each prize-awarding body and submitted to the government for final approval. In this manner, the detailed provisions for the awarding of the prizes could reflect the distinctive character of each field in which awards were to be made. This principle of autonomy was expressed concretely by the deletion of Lindhagen's detailed provisions concerning the ranking of works according to their importance, the nominating system, and the procedures for prize adjudication. Instead, a new paragraph was inserted that set out the principle of special regulations concerning the prizes to be adopted by each prize-awarding institution and promulgated by the government.

Although the changes proposed by Mörner and Tigerstedt were too sweeping to be accepted in their entirety, their proposal nevertheless had a strong influence on the negotiations. This was not so much because the committee majority subscribed to the principle of autonomy, but because the idea of relegating a number of matters to special regulations provided the committee with an easy way of settling conflict. Faced with several divergent opinions on a given issue, the committee typically decided simply not to treat the matter in the statutes, thus following, by a somewhat circuitous route, the Swedish tradition of making consensus the cornerstone of collegial decision making. In some cases such decisions were made in the committee, but for other matters divergent proposals were referred to a small drafting committee composed of Billing, Lindhagen, and Mörner. Here, the presence of Mörner as the sole representative of the prize-awarding institutions must have settled more than one dispute in favor of the Karolinska. As a result, the final version of the statutes that emerged from the Large Nobel Committee after twenty meetings (held between January and April 1899) was a very different document from Lindhagen's draft. The most drastic changes concerned the general principles guiding the process of prize selection: Deleted from the final version were both the definitions of the fields in which the prizes were to be awarded and the ranking of works according to their importance. The discussion of these matters is nevertheless of interest since it brings out the initial conceptions of their task held by those who were to become responsible for awarding the prizes in science and medicine.

In his will, Nobel had referred to "the most important discovery or invention within the field of physics," "the most important chemical discovery or improvement," and "the most important discovery within the domain of physiology or medicine." In the draft statutes, Lindhagen had attempted to clarify Nobel's intentions by inserting descriptive statements that combined Nobel's references to type of work and field. In doing so for physics and chemistry, he enlisted the help of Arrhenius. Together they drafted a statement holding that "physics" and "chemistry" should be taken to mean not only the theoretical parts of these disciplines but also their applications to, and treatment of, problems in related fields. This plan to include interdisciplinary areas undoubtedly reflected Arrhenius's interests in promoting physical chemistry.[37]

The reference to "physiology or medicine" was more problematic since it provided no real clue as to Nobel's intentions. Did he want to promote physiology as a basic field of "science" distinct from the "art" of medicine, or the merging of the two into the medical sciences that was already well under way in the nineteenth century?[38] To try to answer these questions, Lindhagen turned to J. E. Johansson, a lecturer in physiology at the Karolinska who was a close friend of Arrhenius. During a stay in Paris in 1890, when he had done some work in Nobel's laboratory in Sevran, Johansson had gained personal knowledge of Nobel's research interest in the area of medicine. This experience led him to conclude that the reference to "physiology or medicine" in the will should be taken to mean "experimental medicine," that is, "the different sciences which use experimental methods to analyze and, eventually, to cure human disease." The fact that Nobel had placed "physiology" before "medicine" was presented by Johansson as additional support for this interpretation, since this could be taken to mean that in Nobel's opinion physiology represented the most advanced form of experimental medicine.[39]

This interpretation was opposed by the delegates of the Karolinska Institute, whose counterproposal included a statement that work in all the "physiological or medical sciences" (including plant or animal physiology, for instance) could be considered for the prize. When the relevant paragraph was discussed in committee, Johansson presented a detailed memorandum supporting his interpretation of the will. This caused the representatives of the Karolinska to modify their original proposal, restricting their prize area to "the theoretical and practical medical sciences." When the statutes emerged from the drafting committee, however, this statement had been deleted, probably at the instigation of Mörner. As a result, the delegates of the Academy of Sciences requested and obtained the deletion of the part of the paragraph defining the prize areas of physics and chemistry. In the final version of the statutes, the only definition retained was that of literature as comprising not only belles-lettres but also other writings possessing literary qualities (para. 2).

In Nobel's terms, the "discoveries," "inventions," and "improvements" to be rewarded with the prizes should be the "most important" ones. In an attempt to set up standards to guide the prize awarders in this area, Lindhagen had stated as a first principle in his draft that, to merit a prize, a work should be of "the far-reaching importance obviously intended in the will." But he had gone on to suggest that this rule might be relaxed if no major discovery was at hand; prizes could thus be awarded for work that, although not of far-reaching importance, nevertheless implied progress in the area concerned. The prize awarders were also authorized to base their evaluations on the promise that the author of a given work showed for the future. In such a case, it would be necessary for the prize-adjudicating bodies to play an active part in the evaluation, even to seek out promising candidates.

This active mode of evaluation must have seemed novel and alien to the scientists on the committee, not only because it departed from the custom of scientific societies to base rewards mainly on past performance, but also because it contained no specific reference to published works. Hence, the many counterproposals they advanced were made with the aim of bringing the process of prize adjudication into harmony with established practices for recognition and reward of scientific work. In the end, the entire paragraph in which Lindhagen had laid out the general principles that should guide the work of the prize awarders was dismantled and replaced by a single sentence stating that only published works could be considered for the prize (para. 3). Nevertheless, it was necessary to retain the stipulation that, to be awarded a prize, work should be of "the far-reaching importance obviously intended in the will," since this opened up the possibility that if no work met this standard in any given year, the prize could be reserved. This principle, which had been agreed on in the unofficial negotiations, was formalized in the statutes through the provision that the money from reserved prizes would constitute special funds, belonging to the different prize awarders, and that the proceeds from such funds would be used to promote the same purposes as those intended by the prizes but through other means. Elaborate rules were also drawn up so as to ensure that the decision to reserve a prize would not be taken hastily (para. 5).

The decision-making machinery

In contrast to the general principles discussed above, the elaborate machinery for eliciting proposals for the prizes, for evaluating these, and for making the final prize selections fared better, since many of the provisions included in Lindhagen's draft found their way, albeit in modified form, into the statutes. In these very difficult negotiations, the greatest divergences of opinion occurred with respect to the nominating system and the Nobel Institute. In the following, the discussion of the nominating system will be given primary attention, first, because the rule (common to all the prize-awarding institutions) that foreigners be invited to nominate candidates for the prizes was an important element in making the prizes truly international, and second, because the main influence over the selection of prizewinners in physics and chemistry – outside the Nobel committees – came to lie with the nominators.

In Lindhagen's draft, Arrhenius's original idea to invite proposals from members of the prize-awarding institutions as well as from representatives of Swedish institutions of higher education had been expanded to include members of foreign universities and academies. This proposal elicited two counterproposals. The one put forth by Mörner advocated the largest degree of openness to the outside world. His proposal provided for four categories of nominators: (1) members of the prize-awarding institutions; (2) faculty members of Swedish universities and other institutions of higher learning; (3) members of at least one corresponding institution in each of the more important countries of the "civilized world"; and (4) those private individuals whom the prize awarder would see fit to select. By contrast, in the counterproposal advanced by Hasselberg, the appropriate Nobel committee was assigned the task of seeking out worthy candidates and of proposing one of these to the prize-awarding institutions. The committee would have the right, however, to invite a few foreign experts of "high scientific rank" to participate in its work. Between the two poles represented by these schemes there appeared a number of other proposals, most of which reflected some form of compromise.[40]

Hasselberg's proposal was strongly criticized by Arrhenius who, although not a member of the Large Nobel Committee, was closely following its deliberations and making his opinions known to both Lindhagen and Nilson. He tried to win Nilson over to the system proposed by Mörner since, in his opinion, the Hasselberg scheme bore an unfortunate resemblance to the *sakkunnighetsinstitution* that had been used to try to block Arrhenius's appointment to a chair at the Högskola. His views on this system as "strongly discredited" and "yielding particularly poor results" were undoubtedly colored by his personal experiences, but he also warned that Hasselberg's scheme could have unfortunate consequences for the prize selection. "Since the experts would without doubt be sought among the old guard who have neither the time nor the inclination to be concerned with recent developments," he wrote to Lindhagen, "such a scheme would most certainly rule out support of the newer tendencies that Nobel evidently had in mind."[41] Arrhenius need not have feared that Hasselberg's proposal would be adopted, however, since the majority of committee members favored a broader involvement of foreign institutions and individuals.

As the committee was unable to reach agreement on the specific forms that this involvement should take, it was decided to include only the general stipulation that Swedish and foreign representatives of the different subject areas involved be authorized to propose candidates (para. 7). The discussion about these matters had been useful, however, in that it pointed the way to the compromise solution that would eventually be included in the special regulations. This was the system whereby certain categories of individuals were given permanent nominating rights while others were invited to submit proposals only for the prize of a given year.

In the machinery for prize selection, the Nobel committees, one for each prize, represented the second stage of decision making. Despite opposition from the delegates of the Karolinska Institute, who considered this a matter for the special regulations, Lindhagen's version of this paragraph survived the negotiations almost intact. The final version of the statutes thus included the stipulation that membership of the prize-awarding institution was not a prerequisite for election to Nobel committees and that foreigners, too, could serve on the committees (para. 6). It is interesting to note that neither the general stipulation in the statutes that the committees "make suggestions with reference to the award" nor the elaboration on this rule included in the special regulations drawn up for the prizes in the Academy of Sciences (para. 7) presaged the commanding role that the Nobel Committees for Physics and Chemistry were to assume in the selection of prize-winners (see Chapter 6).

In the process of decision making envisaged by Lindhagen, the Nobel Institute would function alongside the committees. It would assist them in evaluating proposals for the prize but not make its own recommendations. Furthermore, it would "create and maintain capability for the accomplishment of the task of awarding the prizes." The protracted debates that arose over the Nobel Institute in the committee did not concern so much its role in the process of evaluation but rather its organizational status. The key issue was whether there should be a single institute organized in different sections, as suggested in the draft statutes, or separate institutes placed under the authority of each prize awarder, as stated in the Karolinska counterproposal. Given the range of opinions on this matter, it is not surprising that the final decision was a compromise solution that contained elements of all the different positions argued in the committee. As outlined in paras. 11–13, each prize awarder would decide whether or not to create a Nobel Institute. Once created, the institute would be managed by the prize awarder in question while remaining the property of the Nobel Foundation. Each institute should remain separate from its managing institution, however, in terms of its financial assets and its personnel. Furthermore, prior approval by the king-in-council was required for joint appointments at a Nobel Institute and any other institution of higher learning. All in all, these provisions spelled the end of the hopes for a big Nobel Institute closely associated with the Högskola.

In the decision-making machinery described in the draft statutes, the final selection of prizewinners in science and medicine was placed in the hands of members of the Academy of Sciences meeting in plenary session and of regular members of the faculty of the Karolinska meeting as a collegial body. In a long paragraph, Lindhagen had proposed detailed rules for the voting and other procedures that should be followed when selecting the prizewinners. In the committee deliberations, these rules were all eliminated on the grounds that differences in the character and composition of the prize-awarding bodies rendered it impossible to draw up rules that would apply uniformly. Only the part of the paragraph stating that prize decisions could not be appealed and that the deliberations should not be recorded or otherwise made known was retained (para. 10). The prize awarders were particularly insistent that these rules be included since it would make their difficult task somewhat easier. It was also felt that the prestige of the prizes would be more secure if one blocked access to materials that might bring adverse publicity.[42] In the special regulations for the prizes in the Academy of Sciences, the secrecy provision was extended to cover not only the final selection of prizewinners but also the deliberations of the Nobel committees (para. 8).

The special regulations

When the Large Nobel Committee had decided on the final version of the statutes on April 27, 1899, this body ceased to exist. Prior to their promulgation by the government, the statutes were first sent to the five prize-awarding institutions for approval and then submitted to the attorney general and the office of the chancellor of the universities for opinions. After the statutes had been accepted by the different institutions, the latter were invited by the government, early in 1900, to submit proposals for the special regulations. In the Academy of Sciences, the elaboration of a proposal was entrusted to a committee composed of the secretary and four other members, among them Hasselberg and Pettersson. The process of drawing up the special regulations turned out to be complex, mainly because of the controversy surrounding the nominating system.

In the Large Nobel Committee, the Academy delegates had held very different opinions on this matter, and these were now carried into the Academy itself. By contrast, at the Karolinska the system that Mörner had advocated in the committee was incorporated into the special regulations and unanimously approved by the faculty. In its final form, this system incorporated the idea advanced in the Large Nobel Committee that certain categories of individuals have permanent nominating rights while others be invited to propose candidates for the prize of a given year. The one accepted by the Academy after lengthy discussion was more restricted in that only Swedish and foreign members of the Academy, members of the Nobel Committees for Physics and Chemistry, and Nobel laureates would have the right to nominate. The final vote on this matter was extremely close, since there was considerable support in the Academy for the less restricted system proposed by the committee preparing the special regulations. This latter proposal included members of Swedish and foreign universities among the groups of nominators.[43]

On June 29, 1900, the statutes and special regulations were discussed and approved by the king meeting with his cabinet. It is somewhat ironic, considering the lengthy debates and the numerous votes on the nominating system in the Academy, that the king-in-council saw fit to change the special regulations of the Academy, bringing the rules concerning nominating rights into harmony with those adopted by the Karolinska Institute (para. 1). This action was justified, according to the minutes, by the close vote with which the Academy proposal had been adopted and by the opinion that the system for eliciting nominations was too restricted. A few other modifications, mainly of a technical nature, were also introduced at this time.[44]

The promulgation by the Swedish government of the statutes

concluded the work, begun early in 1897, of laying the foundation for the Nobel institution. To sum up the results of these labors: They brought into existence an organizational machinery for selecting prizewinners, but they did not produce any specific rules concerning the works, nor even the fields, that would be honored by the prizes. This latter shortcoming does not in any way diminish the considerable achievement of Sohlman and Lindhagen, in particular. With the aid of the Swedish government, they had transformed Nobel's vague ideas into a working structure, despite the objections of Nobel's relatives and the vacillations of the future prize awarders. Although their initial hopes that the support of the Nobel family would not have to be bought had been thwarted, their realism in coming to terms with their demands prevented the will from becoming hopelessly embroiled in controversy. There is no doubt, however, that in purely financial terms the prize-awarding institutions were the main beneficiaries. For the Academy of Sciences, for example, the statutes provided for an initial grant of 600,000 crowns for the establishment of its Nobel Institute, the right to the savings accruing from the money set aside for expenses, as well as to the proceeds of the special funds.

It has been shown how the prize awarders resisted provisions in the statutes that would have restricted their autonomy with respect to the choice of prizewinners. That neither the Karolinska Institute nor the Academy of Sciences wanted to have its prize areas defined a priori is demonstrated not only by their opposition to the inclusion of provisions to this effect in the statutes, but also by the absence of any such statement in the special regulations of either institution. These matters were left, then, to be confronted ad hoc in each successive prize decision. Yet the prize-awarding institutions were not left at total liberty in this matter. The machinery to propose candidates constituted an important mechanism for restricting the prize selection to certain areas and types of work, by bringing the results of the evaluations carried out within national and international scientific communities to the attention of the committees. The juxtaposition of these two modes of evaluation – the judgments of the nominators and those of members of the Nobel Committees for Physics and Chemistry – as well as the weight they carried in the prize decisions of the early years is the subject of the following chapters.

4

An overview of the nominating system and its influence on the prize decisions

The system for selecting prizewinners in physics and chemistry laid down in the statutes was put into practice in September 1900 when the Royal Academy of Sciences met to elect the members of its Nobel Committees for Physics and Chemistry. In the physics committee, the Uppsala group of experimentalists gained the majority of seats through the election of Thalén, Ångström, and Hasselberg. Also elected were H. H. Hildebrandsson, professor of meteorology at Uppsala, and Arrhenius. The chemistry committee was made up of the two professors of chemistry at Uppsala, Cleve and Widman, and three professors from Stockholm institutions, Pettersson (the Högskola), P. Klason (the Institute of Technology), and H. G. Söderbaum (the Academy of Agriculture).[1]

One of the first steps for the newly elected committees was to invite nominations for the prizes.[2] The nominating system performed two important functions with respect to the selection of prizewinners: (1) It provided the committees with an indication of the degree of support that each candidate enjoyed in national and international scientific communities; (2) To the extent that the committees based their selections on such support, the overall number of candidates was narrowed down to a "short list" of serious contenders for the prizes.

It seems reasonable to assume that it would be easier for the committees to base their recommendations on nominators' "votes"[3] when these clustered around the name of a candidate in a given year or when the same candidate was repeatedly nominated over a number of years. The examination of such clustering of nominators' opinions is a necessary first step in considering the influence of the nominating system on the recommendations of the committee as well as on the award decisions of the Academy. The committees were under no *obligation*, however, to base their recommendations on nominators' votes.[4] For example, in 1903, when the physics committee recommended the prizewinners of that year (Becquerel and the Curies), it did so by contrasting its own "purely objective evaluation" of the candidates with the "not always objective factors" underlying the nominations.[5] The following discussion of the effect of the nominating system on committee recommendations therefore addresses two questions: Was there a clustering of opinion among the nominators? and Did the committees choose to take this into account?

In addition to guiding the committees and the Academy in their choices, inviting nominations from Swedish and foreign scientists also gave the latter a participatory interest in the awarding of the prizes. The institution was thus provided with ready-made constituencies of the scientists who made the most frequent use of their right to nominate candidates for the prizes. Later in this chapter statistical data will be used to assess the influence of the statutory categories of nominators over the prize decisions.

Another way of looking at the constituencies that formed around the institution is to consider them in terms of the national origins of nominators and nominees. Did the nominating system mainly function along national lines and, if so, what was the relative weight of different national groups? Or was it international in scope in the sense of candidates receiving significant support from other than their own countrymen? These questions will be addressed in the final section of this chapter.

THE DISTRIBUTION OF NOMINATIONS AND THEIR INFLUENCE ON THE PRIZE DECISIONS

In the period 1901–1915, eighty-three candidates were formally considered for the prizes in physics alone, seventy-one for those in chemistry, and fifteen in both disciplines. Of these, nineteen were selected to receive the prize in physics, fifteen in chemistry, and one (Marie Curie) in both physics and chemistry. The candidates were put forward by some 400 nominators. Since many of these made multiple nominations, the number of nominations was considerably larger than the number of nominators: In all, around 600 nominations in physics and 470 in chemistry were accepted as valid by the two committees.[6] The higher number of candidates and nominations in physics is explained by the more

Table 4.1. *Nobel prize nominations received by candidates in physics and chemistry, 1901–1915*

Number of nominations received	Percentage of candidates (N in parentheses) Physics	Chemistry
1–2	45	55
	(44)	(47)
3–5	21	17
	(20)	(14)
6–10	19	12
	(19)	(10)
11–19	8	8
	(8)	(7)
20 or more	7	8
	(7)	(7)
Total:	100	100
	(98)	(85)

Sources: Nobelprotokoll, KVA; Protokoll, NK, fysik, NK, kemi, 1901–1915.

frequent proposals for divided prizes in that discipline. That this was less often the case in chemistry may have been due to the nominators taking their cues from previous decisions that hardly ever involved divided prizes.

Analysis of the weight that the nominating system carried in the prize decisions is a two-step process. First, it is necessary to establish whether or not the nominations represented clusters of opinion in favor of a given candidate; and second, we must examine the extent to which this type of validation of a candidacy was reflected in committee recommendations and final prize decisions. Overall, the nominating system can hardly be held to have functioned on the basis of majority votes, for the mean number of nominations received by the prizewinners during the period 1901–1915 was quite low: 4.3 in physics and 6.6 in chemistry. This was not significantly different from the mean number of nominations received by all candidates: 6.2 in physics and 5.7 in chemistry.

As indicated in Table 4.1, these figures cover a considerable *range* of nominations received by individual candidates. Slightly less than half (45 percent) of the candidates in physics, and slightly more than half (55 percent) of those in chemistry did not, in fact, receive more than two nominations. The large number of candidates in this category indicates a considerable divergence of opinion among the nominators. Among the candidates receiving more than two nominations, the largest portion received from three to ten votes (40 percent in physics and 29 percent in chemistry). Only around 15 percent of the candidates in both fields received more than ten nominations. It should be noted that the two scientists who received the largest number of votes in absolute terms (fifty-four and forty-nine, respectively) were both candidates in physics (Planck and Poincaré), whereas the two at the top of the list in chemistry (Nernst and Moissan) only received forty-one and thirty-six votes, respectively.

The impact of the nominating system on the prize decisions was quite naturally conditioned by this considerable range of opinion among the nominators. As indicated by Figure 4.1, in both physics and chemistry the prizewinner was the most nominated in only seven of the fifteen years; in chemistry, the majority of these prizewinners were found in the category of candidates receiving ten nominations or more. In most of the other years, the most nominated candidate received less than ten nominations. Partly for this reason, but also because of other circumstances described below, some laureates won their prizes on the basis of a very small number of nominations. To mention some of the more extreme cases: Dalén, the Swedish inventor, received the physics prize of 1912 on the basis of one nomination and none before then; Marie Curie had a total of three nominations for the two years she won her prizes and only two before then; von Laue received his Nobel prize in 1914 – the first year he was nominated – on the basis of two nominations.

In the years 1901–1906, committee recommendations and final decisions were generally based on the clustering of nominators' opinions around certain candidates. This was not surprising given that in the early years nominators and committee members alike could draw on the pool of candidates whose work, most of which dated from the 1890s, was sufficiently well known and accepted to be almost immediately put up for award (e.g., the discoveries of X-rays, radioactivity, the inert gases in the atmosphere, and the electron). A closer examination of Figure 4.1 for the years

1901–1906 reveals the two ways in which nominators' opinions were reflected in the recommendations: by the committee recommending and the Academy choosing the candidate(s) who had received the most nominations in that particular year, and by their taking the candidate(s) who had accumulated the most nominations over several years.[7] It appears, in fact, that the nominations carried more weight with the committees when nominators' votes in favor of a given candidate had time to accumulate. Or, put another way, in this early period, a war of attrition on the part of the nominators would be more effective than a *Blitzkrieg*. This is borne out further by the fact that in the years when the candidate with the highest number of accumulated nominations was not the prizewinner, he would most often be chosen in the year(s) immediately following. Thus, the effect of the convergence of opinion among the nominators in the years 1901–1906 made itself felt until 1909 when the last set of candidates (Marconi and Ostwald), whose nominations had accumulated almost from the beginning, received their prizes.

From 1907 until the end of the period under study, prize decisions in both physics and chemistry were only very infrequently based on nominators' support for given candidacies either because there was no such clear-cut support or because the committees chose to disregard majority votes among the nominators. The former situation was characteristic of chemistry, the latter of physics. That the physics committee so often acted in accordance with its 1903 statement quoted earlier and dispensed with the nominators' opinions was mostly due to an upsurge in "problem candidacies" after 1907. Such a candidacy might be defined, admittedly in a rather arbitrary manner, as one that had accumulated more than ten nominations without resulting in a prize award during the period studied here. There were eight such candidacies in physics, five in chemistry. More important, the ones in physics accumulated a considerably higher number of nominations, as illustrated by the figures cited earlier for Planck and Poincaré.

It would be too simple, despite the prominent position of Planck among the problem candidates, to attribute differences between the nominators' opinions and those of the physics committee members chiefly to upheavals in the discipline caused by the revolutionary view of matter implied by quantum theory. With the exception of Planck's, problem candidacies arose, in fact, less fre-

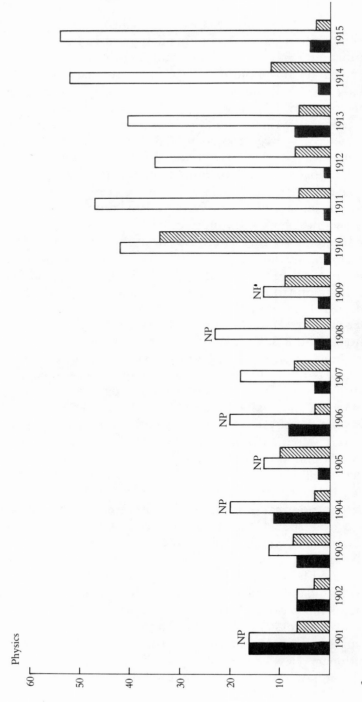

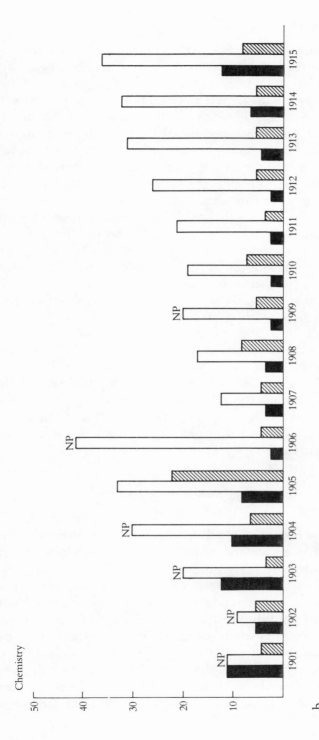

Chemistry

Fig. 4.1 Number of nominations received by the prizewinner of each year as compared with the candidate receiving the highest number of accumulated nominations (sometimes the same person) and the most nominated non-prize-winning candidate: (a) physics, 1901–1915; (b) chemistry, 1901–1915. *Key: solid bar = Nobel prizewinner; open bar = most accumulated nominations (NP = Nobel prizewinner); shaded bar = most nominated non-prizewinner.*

quently over questions concerning the validity of theory than they did from the narrow definition of physics as meaning primarily experimental physics, and from the way the demarcation lines were tightly drawn with respect to new specialties that were part physics, part chemistry. The divergences of opinion between the nominators and the committee members over the inclusion of mathematical and theoretical physics under the rubric of physics are most in evidence in the refusal of the committee to recommend Poincaré for the prize despite the campaign that Mittag-Leffler conducted in support of his candidacy in 1910, and that resulted in his being the most nominated candidate of any year (see Chapter 5). Other problem candidacies were created by the reluctance of the committee to reward work in interdisciplinary specialties (for example, physical chemistry in the case of Arrhenius and radio-activity in that of Rutherford).

The situation in chemistry was different, since, from 1907 to 1914 there was no clustering of nominators' opinions. The most nominated candidates, and also the prizewinners, were most fre-quently those who had received less than ten votes. The one ex-ception here was Nernst, the major problem candidate during the period, since he was among the most nominated in four of the eight years and also had the highest number of accumulated nom-inations (forty-one). His candidacy did not result in an award decision until 1921,[8] however, mainly because of strong opposition from Arrhenius (see Chapter 5).

THE PARTICIPATION AND INFLUENCE OF
DIFFERENT GROUPS OF NOMINATORS

In Table 4.1 and Figure 4.1, each nomination is counted as having a value of one. As mentioned earlier, however, it can be assumed, at least hypothetically, that nominations would carry different weights depending on the *category* of nominator and, ultimately, on the reputation and standing of individual nominators. As a result of statutory provisions that endowed some groups of nom-inators with permanent nominating rights and others with the right to nominate only in a given year, the former had, at least in principle, more influence than the latter. Whether or not this was so in practice depended on the degree of their active participation in the nominating process. These groups of active nominators also formed the initial constituencies of the Nobel prize institution in the broader sense.

The special regulations for the awarding of the prizes in physics and chemistry by the Royal Academy of Sciences provided for six categories of nominators. The four categories mentioned first (paras. 1.1–1.4) were given permanent nominating rights. They were (1) Swedish and foreign members of the Royal Academy of Sciences; (2) members of the Nobel Committees for Physics and Chemistry; (3) scientists who had been awarded the Nobel prize by the Academy of Sciences; and (4) permanent and acting professors of physics and chemistry at universities and analogous institutions in Sweden and the other Nordic countries as well as teachers of these subjects at the Stockholm Högskola. Members of the two remaining categories (paras. 1.5–1.6) were named each year by the Academy upon proposition by its Nobel committees. These categories were (5) chairholders in physics and chemistry from at least six foreign universities and (6) an unspecified number of scientists invited to nominate in their individual capacity.

These conditions ensured that those nominating candidates for the prizes would, in most cases, hold professorial appointments. While this was a requisite for categories 4 and 5, in the Academy of Sciences, too, most Swedish and foreign members were drawn from the upper echelons of academe. At least in theory, the committees were provided with a means for soliciting opinions and expertise from individuals with a more varied background through the use of category 6 (specially invited scientists). However, the nominators in this category were as likely to be university professors as those in any other.

Although the overall group of nominators represented, perhaps in a more abstract than real sense, an "international republic of professors," it was composed neither exclusively of physicists and chemists nor only of scientists who were active in research. On this latter point, it should be remembered that in most countries professorships, because of the heavy teaching and administrative duties they entailed, often represented a considerable hindrance to active research careers. On the former point, the names and affiliations of some of the more active nominators, all members of the Academy of Sciences, testify to the fact that the nominating system was not restricted to the fields in which the prizes were awarded. Among the Scandinavians, for example, there were

W. C. Brögger (professor of mineralogy and geology at the University of Oslo) and G. Mittag-Leffler and E. Phragmén (both professors of mathematics at the Stockholm Högskola); among the Germans, V. Czerny (professor of surgery at the University of Heidelberg), F. von Hefner-Alteneck (an engineer), and H. Brunner (professor of historical jurisprudence at the University of Berlin); among the French, C. Bouchard (professor of pathology at the Paris Faculty of Medicine) and P. Painlevé (professor of mathematics at the Ecole Polytechnique); among the British, Lord Avebury (banker, member of Parliament, and a dedicated amateur naturalist as well as the author of many books of popular science) and Sir W. H. White (director of naval construction and assistant controller of the Navy).

The participation of each statutory group of nominators and their relative weight in the nominating process are presented in summary form in Table 4.2. As might be expected, the "activity ratios" (computed for each category as the number of nominations divided by the number of nominators) are highest for the categories of permanent nominators (Categories 1–4). These also accounted for more than half of the nominations received by all candidates in the two fields (second column group). Here, *foreign* members of the Academy of Sciences and Nobel laureates supplied the bulk of the nominations.[9] Since so many of the laureates were also foreign members of the Academy, it has been necessary to combine these two categories (Categories 1 and 3). Before the first awards were made, in fact, the Academy counted among its foreign members five scientists (von Baeyer, van't Hoff, Rayleigh, Ramsay, and Röntgen) who were subsequently to receive the prize. During the period 1901–1915, nine more laureates (Marie Curie, Fischer, Lenard, Marconi, Michelson, Ostwald, Rutherford, T. W. Richards, and J. J. Thomson) were elected members of the Academy. In some cases, notably those of Lenard, Michelson, and Ostwald, election to the Academy took place shortly before receipt of the prize and may thus have provided a means of testing the standing of the candidate among the membership at large.

As for the nominators in the two categories (5 and 6) who were invited to propose candidates for the prizes of a given year, their combined activity ratios (1.9 in physics and 1.5 in chemistry) were, almost by definition, lower than those of the nominators in the permanent categories. As indicated in the second column group

Table 4.2. *Distribution of Nobel prize nominations by category of nominator, 1901–1915*

Category of nominator	Activity ratio[a]		Percentage of nominations received by all candidates (N in parentheses)		Percentage of nominations received by Nobel prize-winners (N in parentheses)	
	Physics	Chemistry	Physics	Chemistry	Physics	Chemistry
Members of the Royal Academy of Sciences and Nobel laureates (cats. 1 and 3)	3.3	3.5	50 (307)	37 (178)	51 (44)	29 (29)
Members of the Nobel Committees for Physics and Chemistry (cat. 2)	3	4.7	4 (24)	9 (42)	11 (9)	18 (19)
Chairholders in physics and chemistry at Nordic universities (cat. 4)	4.1	3.3	6 (37)	10 (46)	4 (3)	12 (12)
Chairholders in physics at non-Nordic universities (cat. 5)	1.4	1.5	26 (156)	40 (188)	20 (17)	34 (35)
Scientists invited in their individual capacity (cat. 6)	3.6	1.4	14 (83)	4 (20)	14 (12)	7 (7)
Total:			100 (607)	100 (474)	100 (85)	100 (102)

[a] The "activity ratio" is defined as the number of nominations divided by the number of nominators in each category.
Sources: Nobelprotokoll, KVA; Protokoll, NK, fysik, NK, kemi, 1901–1915.

of Table 4.2, chairholders in physics and chemistry at foreign universities (category 5) contributed a significant proportion of the nominations received by all candidates – 26 percent in physics and 40 percent in chemistry – whereas the proportion of specially invited scientists (category 6) was more limited: 14 percent in physics and 4 percent in chemistry. The difference between physics and chemistry in category 6 is explained by the restricted use made of this category by the chemistry committee, where such invitations were not made except in the years 1904–1907 and 1912–1914, and then only to a few chemists. By contrast, in physics from 1903 onward, invitations were extended to a varying number of individual physicists (ranging from eight to eighteen), between one-third and two-thirds of whom would normally respond. Whereas the list of universities from which the chairholders were invited to nominate underwent an almost complete revision each year (presumably to ensure the "just representation of the various countries and their respective seats of learning" called for in the special regulations), that of the physicists invited in their individual capacity was only partly redrawn. This had the effect of creating a nucleus of semipermanent nominators in this category (e.g., N. Lockyer, O. Lodge, M. Planck, G. Quincke, E. Riecke, and E. Warburg) whose participatory interest in the institution is reflected in their high activity ratio.

When one moves on to consider the third column group, the proportion of nominations received by Nobel prizewinners as compared with those going to all candidates, it becomes clear that influence over prize decisions was not always proportional to the numerical significance of each group of nominators.

Not surprisingly, the nominations of members of the Nobel Committees for Physics and Chemistry, numerically the smallest category, carried the most weight since they were the only category of nominators that had the official means to implement their preferences through recommendations to the Academy. The practice of exerting the influence of the committee over the prize selection at the nominating stage was begun in physics in 1901. At a meeting in early January, the members unanimously decided to nominate Röntgen and Lenard for the first year's prize. At that date only one other nomination was at hand, and this by a nominator who had proposed himself for the prize.[10] In subsequent years, nominations by members of the physics committee were often made to ensure that a prospective prizewinner would not be missing from the list of candidates in the year that his "number" might be coming up.[11]

By contrast, in chemistry, committee members used their nominating rights to signal their preferences early on in the process of prize selection. Until 1907, the votes of committee members were almost always part of the consensus among the nominators that carried over into Nobel prize awards. From 1907 to 1915 the nominations of committee members show the same dispersion among different candidates as those of the other nominators, thus adding to the committee's difficulty in arriving at a consensual decision (see Chapter 6). It was also during this time that Arrhenius, the only member of the physics committee who made nominations in chemistry, came to the rescue of the committee by proposing candidates who turned out to be acceptable to the majority of its members. In the years 1908, 1909, 1911, and 1913 Arrhenius, together with a mere handful of other nominators, picked the prizewinners in chemistry, namely Rutherford, Ostwald, Marie Curie, and Werner. This record will receive further attention in the following chapter.

The two other numerically insignificant categories where the nominations carried proportionally greater weight than indicated by the share received by all candidates are both found in chemistry. In the case of chairholders at Nordic universities (category 4), this was due almost entirely to a group of Norwegian nominators (H. Goldschmidt, T. Hjortdahl, and J. Vogt), all professors at the University of Oslo who, acting in unison, and perhaps even taking their cues from members of the chemistry committee, frequently joined in creating the consensus of opinion underlying the prize decisions of the early years. As for the category of specially invited scientists in chemistry (category 6), it did not contain the same nucleus of semipermanent nominators as that in physics, but those invited to nominate nevertheless seem to have had a flair for picking prizewinners.

Although it is not directly apparent in Table 4.2, the category of specially invited nominators in physics nevertheless included the one nominator – Emil Warburg – whose record in picking prizewinners almost matched that of Arrhenius. Acting in both physics and chemistry, Arrhenius nominated eight prizewinners during the period, whereas Warburg nominated six in physics alone (Röntgen, Rayleigh, J. J. Thomson, Wien, Kamerling Onnes, and von Laue). Warburg's remarkable performance was due to a number of factors. The most important were that he occupied a central position in German physics by successively holding the chair of experimental physics at the University of Berlin (1895–1905) and the directorship of the Physikalisch Technische Reichsanstalt at Charlottenburg (1905–1922), positions in which he trained many of the leading experimental and theoretical physicists in Germany; he knew personally and frequently visited the Swedish physicists on the committee; and while he worked in many of the areas in which Nobel prizes were awarded, notably electrical conduction in gases and experiments designed to test empirically the radiation laws formulated by Wien and Planck, he was (and this is perhaps the most important factor) never himself a candidate for the prize.[12]

Numerically, the most significant groups of nominators were categories 1 and 3 combined and category 5, (that is, members of the Academy of Sciences, Nobel laureates, and chairholders at foreign universities). Among them, they made 71 percent of the nominations received by Nobel prizewinners in physics and 63 percent of those in chemistry. Since this was the same proportion that all candidates received from the above-mentioned groups, it meant that each vote has to be counted with a value of one or, sometimes, less. The latter (i.e., votes carrying a value of less than one) was most apparent for the group of Nobel laureates who were not members of the Academy of Sciences. When distinguished from the combined categories 1 and 3 in Table 4.2, their nominations represent 2 percent of those received by prizewinners in both physics and chemistry as compared with 7 and 5 percent, respectively, of the ones received by all candidates. By contrast, the nominations of the Nobel laureates who were *also* members of the Academy show roughly the same proportionality as the combined categories 1 and 3, although the weight of their votes is still less than those of Academy members who were not laureates.

If it is true, as has been claimed by some laureates in recent times, that their opinions strongly influence the prize selections, this does not seem to have been the case in the early years of the institution.[13] Of course, during this time Nobel laureates constituted only a small group. Many of them were in contact with the Swedish scientists who served on the Nobel committees. It is therefore possible that they made their opinions known informally rather than through nominations. That the foreign members of the Academy carried so much weight in the nominations system is an important finding, since it points up how closely the Nobel prizes were linked to exchanges of rewards and honors between prominent scientists, who were also often members of academies or leading scientific societies in their own countries.

NATIONALISM AND INTERNATIONALISM IN THE NOMINATING SYSTEM

The distribution of nominators and nominees among different countries permits one to examine whether the nominating system functioned on an international basis or mainly reflected the validation of work within national scientific communities.

Of course, the term *international* cannot be used in its present-day sense, since all the candidates represented either European countries or North America. That the prize selections would be restricted to what was known as the "civilized world" was indicated by the statutory provision (para. 8) that stated that it was not mandatory for the prize awarders to consider works in other than the Scandinavian languages, English, French, German, or Latin if doing so would involve "a great expenditure of time and money." There is no evidence of this provision having been invoked; even when candidates worked in other languages (Russian, Polish, or Italian, for example), their publications were often available in one or another of the "major" languages.

The relative weight of different national groups in the nominating system is given in Table 4.3, which shows the distribution of candidates, nominators, and prizewinners among the countries or groups of countries participating in the nominating process.[14] Clearly, the small number of prizewinners from each country does not permit one to calculate the percentage distribution of these or to make statements comparing the number of prizewinners with the number of nominators or candidates in the different countries. Even with a larger number of prizewinners, however, the many subjective factors that influenced the nominations and the deliberations of the committees would make it hazardous, to say the least, to venture into such comparisons.

That the weight of German science in the nominating system

Table 4.3 *Distribution of Nobel prize nominators, candidates, and winners by country, 1901–1915*

Country	Physics – percentage (N in parentheses)			Chemistry – percentage (N in parentheses)		
	Candidates	Nominators	Prizewinners[a]	Candidates	Nominators	Prizewinners[a]
Germany	28 (27)	25 (51)	(5)	36 (31)	37 (68)	(7)
France	24 (23)	16 (33)	(4)	20 (17)	12 (23)	(4)
Great Britain	19 (19)	9 (18)	(5)	10 (9)	6 (10)	(2)
United States	6 (6)	8 (16)	(1)	6 (5)	7 (12)	(1)
Sweden	6 (6)	11 (24)	(1)	5 (4)	8 (14)	(1)
Netherlands	4 (4)	4 (8)	(4)	1 (1)	1 (2)	(0)
Italy	2 (2)	8 (17)	(0)	5 (4)	4 (7)	(0)
Other Nordic countries	5 (5)	4 (8)	(0)	4 (3)	4 (8)	(0)
Central and Eastern Europe[b]	6 (6)	9 (19)	(0)	6 (5)	11 (20)	(0)
Other countries[c]	0 (0)	6 (13)	(0)	7 (6)	10 (19)	(1)
Total:	100 (98)	100 (207)	(20)	100 (85)	100 (183)	(16)

[a] The number of prizewinners is too small to be calculated in percent. [b] Austria, Czechoslovakia, Hungary, Poland, and Russia. [c] Belgium, Canada, Spain, and Switzerland.

Sources: *Nobelprotokoll, KVA; Protokoll, NK, fysik, NK, kemi,* 1901–1915.

was considerable but far from all-pervasive is indicated by the proportion of nominators and candidates from that country: around one-fourth of the nominators and candidates in physics and more than one-third of those in chemistry. The latter figure is hardly surprising given the dominance of Germany in that field. The large proportion of German nominators can be attributed to several factors, among the most important of which are the following: (1) German scholars were particularly well represented in the categories of permanent foreign nominators (categories 1 and 3), which, as has been shown previously, had a high activity ratio;[15] (2) the list of universities where the chairholders in physics and chemistry were invited to nominate would always include one or two German ones in physics, and two or three in chemistry (the fact that these two disciplines were most often represented by a larger number of chairs at German universities than they were at those of other countries represented an additional factor making for a strong German presence); and (3) German physicists in the category of specially invited scientists (category 6) were many times more active than their colleagues from other countries.

Whereas the share of German nominators was roughly proportional to that of German candidates (and also to the number of Nobel prizewinners), in the cases of France and Great Britain the proportion of candidates exceeded that of the nominators. For this situation to occur, it would be necessary not only for a substantial proportion of nominators to prefer candidates from their own country, but also for these candidates to receive the support of nominators in other countries. To consider first the extent to which nominators' preferences followed national lines, the data presented in Table 4.4 reveal that in physics this was most pronounced among the French and British nominators, three-fourths of whose nominations were in favor of their compatriots. While "own-country" nominations for France were artificially inflated by the heavy participation of French nominators in the campaign to secure the prize for Poincaré, the even higher figure shown for nominations in chemistry (83 percent) indicates that voting along national lines was a more basic trait among the French nominators than among those of other countries. By contrast, German nominators were less likely to favor their own countrymen (52 and 57 percent of the nominations in physics and chemistry, respectively, being own-country votes), as were those of the United States,

Table 4.4. *Nobel prize nominations made in favor of "own country" candidates by nominators of major countries, 1901–1915*

Nominator's country	Percentage of nominations (N in parentheses)	
	Physics	Chemistry
Germany	52 (97)	57 (94)
France	76 (80)	83 (53)
Great Britain	75 (39)	48 (14)
United States	48 (16)	57 (17)
Sweden	8 (6)	15 (10)
Netherlands	52 (13)	0
Italy	46 (17)	23 (3)

Sources: Nobelprotokoll, KVA; Protokoll, NK, fysik, NK, kemi, 1901–1915.

Netherlands, and Italy. As behooved the country hosting the Nobel prizes, Swedish nominators only rarely favored their own countrymen, but this may also reflect the absence of valid candidates, particularly in physics, where only 8 percent of the nominations were own-country ones. The higher figure (15 percent) obtained for chemistry resulted of course from the support given Arrhenius's candidacy by Swedish chemists. All in all, Table 4.4 indicates that there was a moderate degree of nationalism among the nominators of the major powers, with the German ones showing particular restraint.

The extent of nationalism exhibited in the votes of the nominators only partially answers the question of whether or not the nominating system functioned along international lines. To shed further light on this matter, it is necessary to consider the data from another viewpoint, one that brings out the constellation of votes *received* by candidates of different countries or groups of countries and the distribution of these votes according to their

Table 4.5. *Nobel prize nomination "chauvinism" index, 1901–1915*

Nominator's country	Physics	Chemistry
Germany	37	33
France	62	71
Great Britain	61	39
United States	44	53
Sweden	5	12
Netherlands	47	− 0.4
Italy	44	20
Other Nordic countries	8	22
Central and Eastern[a] Europe	3	6
Other countries[b]	− 2	33

Note: The "chauvinism index" is constructed as follows (using Germany as an example):

$$\frac{N \text{ Germans nominated by Germans}}{N \text{ German nominators}} -$$

$$\frac{N \text{ Germans nominated by non-Germans}}{N \text{ non-German nominators}} \times 100$$

[a]Austria, Czechoslovakia, Hungary, Poland, and Russia.
[b]Belgium, Canada, Spain, and Switzerland.
Sources: Nobelprotokoll, KVA; Protokoll, NK, fysik, NK, kemi, 1901–1915.

national origins. In the "chauvinism index" presented in Table 4.5, the proportion of own-country nominations have been weighed against those emanating from nominators of other countries.[16] A score of 100 on the index would indicate that all the nominations received by the candidates of a given country were from nominators of that country and none from other ones; an index value of zero (or a negative value) signifies the reverse. The scores on the chauvinism index reflect the relatively high proportion of own-country nominations received by candidates of the major powers, but they also show how this effect was offset by nominations originating in other countries. This was particularly noticeable for German candidates, whose low scores on the index indicate that the contribution of votes from the outside was larger than for the French, British, or American ones. The chauvinism index also reflects the extent to which candidates from minor or peripheral countries had to rely on nominations from other than their own countrymen. This was true particularly for Central and Eastern European countries as well as for the smaller ones lumped together as "other countries." This dependence was due, of course, to the fact that nationals of these countries only rarely figured as nominators for the prizes. The one exception was a group of Swiss chemists whose active support of their own countrymen resulted in the higher score for chemistry in "other countries."

When one charts the national origins of the largest proportion of nominations that each country received from the outside, one finds some distinct patterns that seem to indicate that the nominating system functioned not so much along international or even bilateral lines as it did along unilateral ones (see Fig. 4.2). Given the long-standing ties that existed between Sweden and Germany, it is to be expected that these two countries should form a prominent axis. This is confirmed, particularly in physics, where Sweden provided German candidates with the highest proportion of nominations (48 percent) received from the outside. German nominators reciprocated by providing the same high share of the vote of Swedish candidates, thus making for the only clearly bilateral relationship found among the nominating countries. In chemistry, too, the largest contribution to the German vote came from Swedish nominators but the share it represented was less important than in physics (26 percent), and it was not reciprocated. The largest portion of nominations of Swedish chemists (73 percent) emanated, instead, from nominators in the other Nordic countries.

The most interesting aspect of Figure 4.2, though, lies in the different overall patterns shown for physics and chemistry. In physics, outside the Swedish–German axis, six other countries received the largest share of their nominations from either Sweden or Germany. The other two main "suppliers" of nominations were France and Great Britain (to the Netherlands and Italy, respectively). In chemistry, as already noted, there was no Swedish–German axis; instead, Germany alone supplied the largest proportion of nominations for six other countries. With the exception of Sweden, only two other countries or groups of countries (Great Britain and Central and Eastern Europe) figure as "suppliers" of nominations, but these relationships are somewhat spurious due to the low number of nominations involved.[17] The extraordinary position of German chemistry within the nominating system is

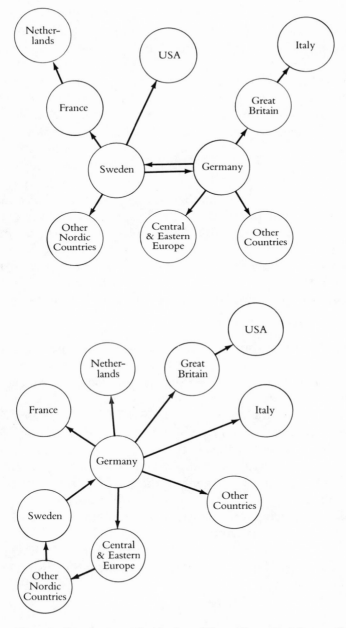

Fig. 4.2. Provenance of majority of "votes" received by candidates in different countries: Top: physics, 1901–1915; Bottom: chemistry, 1901–1915.

perhaps not surprising given the extent to which German universities had functioned as a breeding ground for chemists of many other countries, among them the United States, in the late nineteenth and early twentieth centuries. Whether the German show of support for chemists (and, to some extent, physicists) in other countries resulted from a commitment to internationalism in science, or from a form of scientific imperialism that would have come from favoring the countries that had been "colonized" by German-trained scientists, is of course not possible to ascertain given the absence of knowledge about the motives underlying the nominators' votes.

One might conclude, then, that (1) there was a Swedish–German axis that functioned particularly in physics; (2) Germany and to a lesser degree Sweden occupied key positions in the nominating system in both physics and chemistry, as shown by the large share of votes that their nominators provided for candidates in many other countries; (3) in the case of most of these latter, their nominations were too dispersed to reveal any significant national patterns.

It has been shown how it was only in the first period (i.e., to about 1906) that the nominating system actively guided the committees in their choices. Even during this period, however, the weight of this system did not just depend on majority votes, since considerations of the individuals involved both as candidates and as nominators were certainly important to the committees. In the absence of a consensus among the nominators or when the committees chose to disregard majority votes, such considerations took on even more importance. One of the major factors determining the weight of the nominations was the phenomenon of informal networks, centering on research orientations and schools, that linked the nominators, the candidates, and their Swedish correspondents. This matter will be studied in the next chapter, with particular reference to the role of two prominent Swedish scientists and their networks.

5

Networks at work in the prize selections: Arrhenius and Mittag-Leffler

The two Swedish scientists whose international reputations gave them the most influence over the prize selections were Svante Arrhenius and Gösta Mittag-Leffler. This influence depended mainly on their places in specialty and other networks whose members took an active interest in the awards. The networks gave them their listening posts abroad, and at home they performed the same role for foreign network members. More important, both men took full advantage of their positions in the decision-making machinery set up for the prizes to promote candidates whose choice they felt would enhance the prestige of the new institution.

Of the two, Arrhenius was more directly involved in the internal process of evaluation of candidates in both physics and chemistry – in physics because he was a member of the committee, and in chemistry because his position amounted to an unofficial membership. In view of the important role he played, his predilections and actions will be given primary attention in the present chapter. By comparison, Mittag-Leffler, who was a member neither of the Nobel Committee for Physics nor of the corresponding section in the Academy of Sciences, had to rely on the nominating system to drum up support for his candidates. The campaigns he mounted are of interest because they demonstrate the many difficulties involved in using this system to produce prizewinners.[1]

Both men sought to use the prizes not just to provide recognition for individuals whose achievements they valued, but also for what they considered significant new developments in physics and chemistry, often those on the border of these disciplines. In both these respects, they were more adventurous, and also closer to science in the making, than the majority of the members on the committees, who confined their choices to generally accepted works in such mainstream specialties as experimental physics. It is not surprising, then, that the actions of Arrhenius and Mittag-Leffler gave rise to two major problem candidacies, those of Planck and Poincaré. The obstacles they encountered in charting a new course for the prize selections were mostly of their own making, however, for their personal conflicts meant that they could never act in unison for any candidate.

Arrhenius and Mittag-Leffler also differed from the majority of committee members in the responsibilities that each assumed within and for the institution. Whereas the committee members were faced, as a group, with the burdensome task of selecting prize-winners for each of the fifteen years studied here, Arrhenius and Mittag-Leffler intervened, on their own, in support of a few outstanding individuals. For this reason, their actions, despite their influence on several prize decisions, do not so much exemplify routine decision making about the prizes (see Chapter 6) but show how two strong personalities influenced the institution.

THE CONTRASTING PERSONALITIES AND OUTLOOKS OF ARRHENIUS AND MITTAG-LEFFLER

While Arrhenius and Mittag-Leffler each moved in their distinctive spheres abroad, given the small size of the Swedish scientific community this was not possible at home. Here, the realization of their scientific and personal ambitions was accompanied by frequent clashes between the two. Their contrasting personalities and outlooks were at the root of their mutual distrust.

With his robust health and optimistic outlook on life, Arrhenius was an outgoing personality with an unusual capacity for making friends. Learning of the suicide of Boltzmann, under whom he had studied in Graz in the 1880s, he wrote to Ostwald: "I consider myself at the opposite pole from him [Boltzmann] since no human being can be luckier than I am. My work goes well and I live only for the highest ideals. My health and my living-conditions represent everything I can hope for."[2] While Arrhenius's buoyancy shines through in his letters and personal writings, there was a darker side of his character that made him doubt his ability as a scientist and led him to rely on colleagues whom he perceived as possessing more authority and willpower than he himself could muster. The adulation Arrhenius showed Ostwald early on in their

Svante Arrhenius (1859–1927) in the Physics Institute at the Stockholm Högskola (1901). (Photograph by Anton Blomberg, courtesy of the Stockholm City Museum)

Gösta Mittag-Leffler (1846–1927), Sweden's best-known mathematician, in front of his palatial villa (now the Mittag-Leffler Institute) in Djursholm outside Stockholm (1907). (Photograph by Anton

relationship is a case in point, but this did not preclude a certain amount of tension between the two. Arrhenius's biographers generally agree that the poor reception given his doctoral dissertation presented in Uppsala in 1884, which contained his first version of the theory of electrolytic dissociation, inflicted a wound that never healed.[3] His bitterness toward the members of the jury was directed particularly at Thalén, professor of physics and Arrhenius's future colleague on the Nobel Committee for Physics, who had already refused him access to the Uppsala physics laboratory and now was instrumental in giving his dissertation the low grade (*non sine laude approbatur*) that contributed significantly to his difficulties in launching a scientific career in Uppsala.[4] From then on, he showed an excessive preoccupation with the regard in which his work was held by the Uppsala physicists. Invited to lecture at the Royal Society of London after he had received the Nobel prize, he wrote to Bjerknes about the accolade he had received from British physicists, adding that "it looks as if in the area of cosmical physics as well, Hasselberg, Hildebrandsson and Ångström will have the pleasure of first leering and then heartily congratulating me on my success."[5]

Mittag-Leffler, by contrast, had few of the features that endeared Arrhenius to his friends. In the diary he kept for most of his adult life his bad health is a recurrent feature. From the turn of the century (when he was in his fifties) onward, this increasingly preoccupied him. As he himself recognized, his health problems could be attributed at least partly to a nervous disposition and an overriding ambition to succeed, not only as a mathematician but also in the many related endeavors to which he applied his organizational and other skills. In whichever field he went into action, Mittag-Leffler used all means available to him to promote his various causes. Whether his aim was to secure financial support for the journal *Acta Mathematica* or the Högskola, to promote or block candidates for academic positions, or, later in his life, to influence Sweden's defense and foreign policy, the ingredients of his "campaigns," as he himself called them, were essentially the same. A stream of letters and cables would flow from his desk exhorting and cajoling those he regarded as his friends into supporting his course of action. When necessary, journalists from leading newspapers would be summoned and "instructed," as he put it, in the proper way of handling the matter. In all this he displayed the skills of a public relations man or lobbyist operating on the level of present-day professionals. To him, all his actions were justified by the righteousness of the cause. To many of his contemporaries, though, Mittag-Leffler appeared as nothing less than Machiavellian in what they felt was his self-seeking pursuit of power and influence, and their judgments of his person were correspondingly harsh. In his memoirs, Arrhenius recounts how when he came to the Högskola in 1891, largely through the help of Mittag-Leffler, the latter offered to lend him money. "If I understood him correctly," he writes, "it was a question of buying my support for his viewpoints. I despised him and have never since had reason to change my opinion."[6]

Even without this incident, it is difficult to see how the two could have cooperated, since they were almost exact opposites on so many counts. Although their social origins were similar, Arrhenius considered thrift both a necessity and a virtue, whereas the life style of Mittag-Leffler was lavish in extreme.[7] Arrhenius's simplicity was reflected in his not being particularly interested in rewards and honors, at least not until after his Nobel prize, when even some of his friends felt that the acclaim he had received had gone to his head.[8] For Mittag-Leffler, by contrast, honors were an important means of extending his sphere of influence or, as he himself saw it, of promoting his causes. At the turn of the century, he was the only Swedish scientist to have been elected foreign member of both the Royal Society of London and the French Academy of Sciences, a dual honor that was not extended to Arrhenius until 1911. Characteristically for Mittag-Leffler, the election to the Paris Academy was won only after a vigorous campaign.[9] Politically and ideologically, the two also belonged to opposite camps. Arrhenius generally held liberal reformist views, whereas Mittag-Leffler's stance was on the whole elitist, nationalist, and monarchist. On specific questions, however, he would sometimes hold more advanced opinions than his reformist colleagues, as evidenced by his support of women scientists. It has already been noted how he brought Sonya Kovalevsky to the Högskola, but he also intervened on behalf of Marie Curie on the occasion of both her first Nobel prize in physics (1903) and her subsequent one in chemistry (1911).

Coming as it did in the aftermath of the battles at the Högskola where the "young Turks" led by Pettersson and Arrhenius had successfully fought the old guard led by Mittag-Leffler, it is not surprising that the Nobel prizes should become the scene of further conflicts between Arrhenius and Mittag-Leffler, especially in view of the activist roles both men assumed with respect to the institution. While Arrhenius was involved in drawing up the statutes, Mittag-Leffler played no role at this stage. Although both the "missing prize" in mathematics and the exclusion of the Högskola from Nobel's final will probably angered Mittag-Leffler, he did not show this publicly. He may have taken his revenge privately by spreading the story of how Nobel, who was fifteen years his senior, had lost out in their presumed rivalry over a woman. His official silence may also have been due to his feeling that he could influence the institution from within, primarily by ensuring that, despite the absence of a mathematics prize, mathematical physics would be an eligible category for the physics prize.

Once the annual awarding of the prize commenced, Mittag-Leffler became active in major and minor matters. As the unofficial representative of the Academy of Sciences, he was often the first to send out cables and letters to the prizewinners informing them of the Academy's decision. This was also the occasion for inviting them to the dinner that he and his wife would give in honor of the laureates in their palatial villa in Djursholm, outside Stockholm. These exercises were of minor importance, though, compared to the campaigns he conducted among the nominators, using his network of foreign connections in support of the mathematical physicists whom he felt merited the prize.

These campaigns were conducted with the avowed aim of "opening the doors to theory" in the award of the physics prize.[10] His reverence for mathematics – "the science of pure thought, the science of the sciences"[11] – meant that for him "theory" in physics was represented by a high degree of mathematization. The content of problems in physics was of subsidiary interest; what counted were the advances in pure mathematics wrought by the interaction of mathematics and mathematical physics. In this respect, his views paralleled those of Poincaré, who stated at the end of his life: "It is a remarkable fact that the works of [mathematical] analysts have been all the more productive *for physics*, the more exclusively they have been pursued for their own beauty."[12] For Mittag-Leffler, there were no works in mathematics or mathematical physics exhibiting more clarity and beauty than those of Poincaré. Hence, the chief aim of his campaigns became that of securing the Nobel prize in physics for Poincaré.

To realize his aim, Mittag-Leffler drew on the close relationships he had developed through *Acta Mathematica* with French mathematicians, in particular Darboux and Painlevé, whose interest in the prizes was linked to their positions in the French Academy of Sciences. As the most prominent foreign member of the academy in Sweden, Mittag-Leffler willingly undertook to intervene in favor of French candidacies, not just that of Poincaré which he promoted most actively. That Darboux, as one of the two permanent secretaries of the academy,[13] came to play an important role in these actions was due to the manner in which the academy, alone among major scientific societies, treated the Nobel prize nominations of its members (as well as those of some French scientists who were not members): On several occasions, nominations for the physics prize, in particular, took the form of "corporative nominations" signed not only by officially invited nominators but also by other scientists generally found among the academy membership.[14] The academy's united action in supporting French candidates was a major reason for the high proportion (around 80 percent) of "own-country" nominations noted in Chapter 4.

Arrhenius did not have Mittag-Leffler's single-mindedness with respect to either national or scientific preferences. The way he moved between different specialities and fields throughout his scientific career testifies to his intellectual versatility. The boldness Arrhenius had shown in his own research, first in formulating his theory of electrolytic dissociation as a young man and later in developing his theories in cosmical physics, also made him able and willing to use the Nobel prizes for recognition of work that was in advance of its time. This should not be taken to mean that he was a revolutionary in the sense of questioning the prevalent mechanical world view, which would undergo a profound revision through the work of Einstein and Planck. Having been trained neither in mathematics nor in theoretical physics, it was not possible for him, even if he had been so inclined, to speed the introduction of relativistic or quantum physics into Sweden.

Although Arrhenius's international reputation was founded on his identification with the Ionists, in the period before the First World War he extended his network far beyond this predominantly German-speaking group of physical chemists. After 1903 he enjoyed not only the prestige conferred on him by his Nobel prize but also the freedom from teaching and administrative duties allowed by his position as director of the Nobel Institute for Physical Chemistry. Among his international contacts, those with the British and American scientists were particularly important since they created a counterweight to what might otherwise have been an even stronger German influence over the prizes in physics and chemistry. In Great Britain where, as noted earlier, the Ionists had found a strong supporter in Ramsay, Arrhenius now broadened his contacts to include members of the first generation of atomic physicists, in particular, J. J. Thomson and Rutherford. In 1909, he spent some time at Rutherford's laboratory in Manchester working, as he himself put it, "as a student in the area of radioactivity."[15] In the United States, where Arrhenius had gone in 1904 to lecture at the University of California at Berkeley, he established contacts with the astrophysicists at the Mount Wilson and Lick Observatories, W. W. Campbell and G. E. Hale, and the following year he participated in the Lick Observatory eclipse expedition to Spain. In 1911, he returned to the United States to deliver the Silliman Lectures at Yale University.[16] In addition, he made annual journeys to Germany where he attended scientific meetings and visited colleagues.[17]

While the international contacts of both Arrhenius and Mittag-Leffler gave them considerable leverage in the nominating process, translating this into actual prize awards required mustering support for their candidates in the Nobel committees and the Academy of Sciences. Mittag-Leffler, who was neither a member of the Nobel Committee for Physics nor of the physics section in the Academy, had as his chief supporter on these bodies Hasselberg, whose appointment as Academy physicist he had helped engineer. When Carlheim-Gyllensköld was elected to succeed Ångström as a member of the Nobel physics committee in 1910, that committee acquired a representative of mathematical and theoretical physics and Mittag-Leffler found an ally in his efforts to procure the prize for Poincaré. To bring more substantial troops into battle, however, Mittag-Leffler had to await the plenary meeting of the Academy at which the year's award decisions were to be made. Here, he would have the support of his disciples from the Högskola mathematics department, I. Bendixson, H. von Koch, and E. Phragmén, who made up the majority of the mathematics section in the Academy. Whether or not and on what grounds members from other sections would follow Mittag-Leffler's lead is difficult to say, since, given the secrecy of the deliberations and the votes, the lineup in the Academy can only occasionally be gleaned from letters or Mittag-Leffler's own diary.

Arrhenius's position on the Nobel Committee for Physics was a difficult one for several reasons. First, he did not share the strict empirical convictions of many of the Uppsala physicists on the committee. Second, Arrhenius's atomistic views were probably at odds with the scientific outlook of the Uppsala physicists who were more likely to share the prevalent Swedish attitude described by one observer as "one of pronounced hostility toward atomists and toward atomic theory."[18] Third, until Carlheim-Gyllensköld joined the committee, Arrhenius was the only representative of the Högskola, which, as noted in Chapter 2, was regarded by many Uppsala professors as an upstart institution. In the physics section of the Academy, too, he was initially the Högskola's only representative. Other members of the younger generation at the Högskola, N. Ekholm, H. E. Hamberg and V. Carlheim-Gyllensköld, later joined the section, but they never came to make up the majority of its membership. Still, the Högskola contingent at the Academy led by Arrhenius was sufficiently strong to constitute an opposition force in plenary meetings with which Mittag-Leffler and his supporters had to reckon.

While the Uppsala majority on the physics committee opposed Arrhenius on both personal and scientific grounds, the situation was quite different among the chemists. By the time the Nobel Committee for Chemistry was instituted, the majority of its members had come to recognize the far-reaching implications of Arrhenius's physical theory of solutions for their discipline. It was among the chemists, then, that Arrhenius found the support, which was to manifest itself in the awards for the original triumvirate of Ionists (van't Hoff, Ostwald, and Arrhenius himself) and in highlighting works bearing on atomic theory in the prize decisions.

THE INFLUENCE OF ARRHENIUS

Rewarding the Ionists

There was never any doubt, given Arrhenius's presence on the scene, that the work of the leading Ionists – van't Hoff, Arrhenius and Ostwald – would figure prominently in the discussions about the prize awards of the early years. When these discussions began, more than a decade had elapsed since Arrhenius and van't Hoff had formulated their theories of solutions, and their far-reaching significance was by now well understood. In an article in the *Philosophical Magazine*, for example, Arrhenius enumerated no less than seventeen basic problems in electrochemistry and related fields that had been resolved by application of his theory of electrolytic dissociation, a claim in most parts substantiated through subsequent developments.[19] By the early twentieth century, as Dolby writes, "as a result of the intensive study of electrolytes and solutions triggered by the debate [over the physical theory of solution], physical chemists had exploited the most productive topics and had either settled or postponed the divisive issues."[20] With the perspective that had been gained on the contribution of the pioneers, then, the question was not if the leading Ionists should be rewarded, but when and how this should be done.

The questions of when and how were posed in the first year the prizes were awarded: van't Hoff and Arrhenius were both nominated. Specifically, these questions concerned, on the one hand, in which order the two should receive the prize and, on the other, to which field, physics or chemistry, the work of Arrhenius had contributed the most. As for the question of order, the nominations did not provide much guidance, for although van't Hoff received the largest number of nominations in chemistry in 1901, he was being commended for his overall contribution to theoretical chemistry, organic as well as inorganic. Since, according to Nobel's testament, the award should be given for a specific "chemical discovery or improvement," the committee had to single out the achievement that would form the basis for an award. This process had already begun at the nominating stage when Cleve and Pettersson, the two most influential members of the chemistry committee, proposed Arrhenius and van't Hoff for a joint prize, citing both Arrhenius's papers on the electrolytic theory of dissociation published in the *Proceedings* of the Swedish Academy of Sciences (1884 and 1887) and the memoir by van't Hoff ("Lois de l'équilibre chimique dans l'état dilué, gaseux, ou dissous") published in the same journal in 1886.[21]

By proposing van't Hoff and Arrhenius jointly, Cleve and Pettersson were advancing the most judicious solution to the problem of who should be rewarded first. However, it was not the solution adopted by the committee or by the Academy, for van't Hoff received the chemistry prize in 1901 and Arrhenius in 1903. Although Arrhenius later claimed that he had postulated his theory of ionic dissociation in his ill-fated doctoral dissertation, published in the *Proceedings* (1884), historians of chemistry who have studied the matter agree that his theory only emerged in his paper of 1887. By that time van't Hoff's paper had already appeared in the *Proceedings* (1886) and had furthermore been instrumental in setting Arrhenius on the path that would lead to his theory.[22] These points of chronology would obviously have been important had the issue at hand been one of simultaneous discovery.[23] Instead, it was the significance of the works of the two men in terms of their *complementarity* that was being stressed in the Pettersson–Cleve proposal.

The chief innovation in van't Hoff's paper was the analogy he established between dilute solutions and perfect gases through the use of osmotic pressure (which he found to correspond to gaseous pressure). This made it possible for him to apply thermodynamics to the calculation of the chemical equilibrium of dilute solutions, expressing his general results as $PV = iRT$, where P is the osmotic pressure, i an empirical factor characteristic of a given solution, and R a constant that is the same for solutions and gases (V and T being, of course, volume and temperature). The term i remained the puzzle in the equation, for it was an ad hoc variable that van't Hoff introduced to correct for the deviations from the characteristics of ideal solutions exhibited by electrolytes. It was this puzzle that sparked Arrhenius's imagination and led him to postulate that the dissociation observed in electrolytes did not result in chemically "active" and "inactive" molecules, as he had stated in his thesis, but in the formation of ions. This idea was supported by his finding that the number of ions produced by dissociation corresponds to the coefficient of i. Whatever the specific merits of the work of each man (and they were undeniably great), when viewed in terms of their complementarity it was Arrhenius's idea that established the generality of van't Hoff's earlier results. This point, made clearly in the Pettersson–Cleve proposal, was perhaps most explicitly stated by van't Hoff in his Nobel lecture when he said: "I should not have had the pleasure of giving this lecture if Professor Arrhenius had not succeeded in demonstrating the cause of these exceptions [i.e., of weak electrolytes from the law of ideal solutions]."[24]

Why, then, did Arrhenius not receive the prize ahead of or together with van't Hoff? There may well have been a feeling among members of the Academy that to attribute all or a share of the first year's prize to a Swedish national would be to diminish the international standing of the prizes. Subsequent events also point to the importance of the second question posed above, that is, in which field should Arrhenius receive the prize?

The most likely reason the Academy passed over Arrhenius is that he made it known that he did not want to receive the prize in that year. For reasons cited above, he may not have been sure of having sufficient support in the Academy. He may also have been holding off in the hope of receiving the prize in physics. The latter explanation carries the most weight in view of the strategy that Pettersson devised to bring this about (as discussed below). Also, in 1902 and 1903 Arrhenius withdrew from the prize discussions in the physics committee, giving his own candidacy as the reason.[25] There were several reasons Arrhenius wanted to be awarded the prize in physics, among which the following were the most important. First, although Arrhenius and his colleagues had set out to revolutionize chemical theory, this had been largely accomplished by applying the methods and reasoning of physics (chiefly thermodynamics) to the problem of chemical activity in solutions. Second, since the 1890s Arrhenius's own research had increasingly been devoted to physics, particularly cosmical physics. Third, given his feelings of resentment toward the Uppsala physicists, there could have been no sweeter victory, or revenge, than to be awarded the Nobel prize in spite of his old adversaries.

In 1902 van't Hoff, in his capacity as a former prizewinner, nominated Arrhenius for the physics prize in a proposal that emphasized the physics components of the latter's theory of electrolytic dissociation. In that year, the physics committee was helped out of potential embarrassment through the campaign which Mittag-Leffler launched in favor of H. A. Lorentz. In chemistry, after the first year's celebration of physical chemistry, it was not surprising that the prize was awarded to an organic chemist, Fischer, who was honored for his work on sugar and purine synthesis. By late 1902, the intransigence of the physics committee in reaction to proposals for an undivided prize for Arrhenius must have been sufficiently clear for his supporters to shift strategy and to propose instead that he receive a share of the prize in both physics and chemistry. For the prize of 1903, then, van't Hoff renewed his proposal in physics, but he also nominated Arrhenius in chemistry. In that year Arrhenius received seven nominations in physics and twelve in chemistry.[26]

Early in 1903, the chemistry committee, probably at the instigation of Pettersson, addressed a letter to the physics committee stating that the time for rewarding Arrhenius had come, and that judging by the distribution of nominations, the award could be made in either physics or chemistry. The solution favored by the chemists, though, was that Arrhenius be awarded half of the prize in physics and half of that in chemistry. Awarding Arrhenius a share of the physics prize was justified, the letter stated, because his theory of electrolytic dissociation had its origin in a physical problem, that of the electrical conductivity of solutions, and because it was able to explain a wide range of phenomena in physics, particularly those concerning the ionization of gases and related radiation phenomena (cathode rays, X-rays, and Becquerel rays). The letter suggested that the other half of the chemistry prize should go to Ramsay for his discovery of the inert gases, argon and terrestrial helium. This would require, however, that the physics committee likewise entertain the possibility of awarding the remaining half of its prize to Lord Rayleigh, whose discovery of argon had been made more or less simultaneously with that of Ramsay and that, furthermore, had been announced in a jointly authored paper (1895).[27]

The chemists' letter received a cool reception from the physicists, whose reply stated that Arrhenius's theory of electrolytic dissociation "indisputably . . . has been most consequential for chemistry." Although the physics committee did not deny that Arrhenius's works had been of significance for physics as well, it did not feel that they "were of greater, or even as great significance as others proposed for the physics prize." Since the committee did not want to pass over these other candidates, it could not recommend Arrhenius for the prize. The proposal that the two committees join forces in rewarding Rayleigh and Ramsay would be acceptable to the physics committee only if each committee decided to recommend them each for an undivided prize, that is,

Rayleigh for the physics prize and Ramsay for that in chemistry. With somewhat distorted logic, the physics committee argued that dividing a prize between two persons who jointly had made a discovery in no way diminished its value, whereas attributing half a prize each to Rayleigh and Ramsay and a whole one to Arrhenius would be a slight to the former two.[28]

The reply of the physics committee quashed Arrhenius's hopes that it would recommend him for any part of the physics prize; moreover, it spelled trouble for the chemists. The reply indicated that if they were to recommend to the Academy that Arrhenius and Ramsay share the prize, the proposal would not only appear to be disloyal to the physicists but would also have little chance of succeeding over the physicists' opposition. The choice for the chemists, then, was a stark one: either Ramsay or Arrhenius. There was more at stake, though, than a simple choice between the two, for the candidates favored by the physics committee – those "others" whose work was of greater significance than Arrhenius's – were Becquerel and the Curies. Looking beyond the matter of rewarding Arrhenius, some members of the chemistry committee, Widman and Söderbaum in particular, saw the award of the physics prize for pioneering work in radioactivity as a dangerous precedent, since it could mean that in the future this field would be appropriated by the physicists for their prize. For Widman (writing to Söderbaum), an important reason to support Ramsay in combination with Rayleigh (with or without Arrhenius) was that "in this manner, the physicists and the Academy would be saved from the foolishness of pulling in the *chemistry* couple Curie already this year," thereby indicating the chemistry committee's interest in keeping the whole matter of the Curies' award open.[29]

The first vote on the 1903 prize by the chemistry committee took place in the spring. Widman, Söderbaum, and Pettersson supported Ramsay for the prize, whereas Cleve and Klason favored Arrhenius. Given this split, the committee could do no more than await the decision of the physicists, which would not come until the fall, and meanwhile obtain reports on the leading contenders. The task of arguing for the importance of Arrhenius's theories for physics fell on Pettersson, who must have hoped that the Academy could be persuaded to award a share of the physics prize to Arrhenius. Cleve, strongly in favor of Arrhenius being the sole recipient of the chemistry prize, took on the task of reporting on the significance of his theory for chemistry, while Söderbaum was assigned the report on Ramsay.[30]

When the physics committee met in the fall, it unanimously decided to recommend Becquerel and the Curies for the prize. To prepare the ground for a favorable consideration of the recommendation, Ångström gave a lecture on radium, accompanied by experiments, to the members of the Academy.[31] The committee's recommendation was endorsed by the physics section and later adopted by the Academy as its final award decision. However, there was one important modification. At the request of Cleve and Söderbaum, the Academy decided to commend the Curies, not as the physics committee had proposed for "their discovery of the spontaneously radioactive elements" but, as the final award citation read, for "their joint researches on the radiation phenomena discovered by Professor Henri Becquerel." In making this change the Academy was undoubtedly influenced by Cleve's argument that "the discovery of such a singularly remarkable element as radium might eventually be considered for a Nobel prize in chemistry."[32] (This of course happened in 1911 when Marie Curie received her second prize.) In the chemistry committee, the final vote still showed a three-to-two split in favor of Ramsay. In the chemistry section, however, Arrhenius gained a majority of seven votes as against three for Ramsay.[33] At the plenary meeting of the Academy it was decided, at the request of the physics committee, that exceptionally the Academy would vote on the chemistry prize before moving on to the physics one. This was necessary since the physics committee as well as the section had also made an alternative recommendation for Rayleigh, but only on condition that the chemistry prize be awarded to Ramsay alone.

The draft of Widman's speech to the Academy, preserved among his papers, represents a concise statement of the two main considerations that motivated the committee majority to propose Ramsay for the prize. First, it shows clearly that Widman, and probably also Söderbaum and Pettersson, had not abandoned their earlier proposal for coupling the prizes in physics and chemistry by awarding them jointly to Arrhenius, Rayleigh, and Ramsay. Widman felt obliged to appeal directly to the Academy, since the initial refusal of the physics committee to consider the proposal had meant that it could not be discussed either in the committees or the sections of physics and chemistry. Second, it shows how important it was to him and his colleagues that the Academy delay its decision to present the prize to the Curies, since, in his opinion, even if their work had not yet been the object of proposals for the prize in chemistry, this situation was bound to change in the years to come. He argued that the isolation of radium (accomplished by Marie Curie in 1902) was of greatest significance to chemistry, since it was the most important new chemical element discovered in recent times. Furthermore, the discovery was likely to change chemists' notions of the basic and invariable nature of the elements.[34] Widman's speech notwithstanding – and he was no doubt supported in his stand by Söderbaum and Pettersson – the Academy decided to award the prize in chemistry to Arrhenius. It then awarded the physics prize to Becquerel and the Curies.[35] As shown by the foregoing, this decision was partly the result of the resistance of the majority on the physics committee to having Arrhenius receive a share of their prize.

Now that both van't Hoff and Arrhenius had received prizes, there remained of the original triumvirate of Ionists only Ostwald. In 1904, it seemed as if his award was imminent, for in that year he advanced to the select group of candidates whose work was made the subject of special reports in the chemistry committee. The report (written by Pettersson) mainly served, however, to point up the difficulties raised by Ostwald's candidacy. Although Ostwald had done research, lectured, and written textbooks on most aspects of modern chemistry and had also been instrumental in winning acceptance for the theory of ionic dissociation, it was difficult to pinpoint one particular achievement that could be held up as an important discovery. This was stated in the general report on the candidates prepared for the Academy in 1904: "Were the Nobel prize ... to be awarded for scholarly work promoting the general development of chemistry, there can be no doubt that Ostwald should have been put ahead of other candidates this year."[36] This statement was reiterated in the general reports of 1907 and 1908. As far as the committee was concerned, then, Ostwald's candidacy was put in abeyance in 1904 and was to remain so until the committee reversed its position in 1909, the year he received the prize. In those years, his chances of being awarded the prize appeared to be diminishing, since a decision would have to be based on achievements that were receding into the past. In fact, by 1906 Ostwald was no longer active in chemical research, having left his chair at the University of Leipzig to set himself up as an independent scholar and philosopher and devoting his time to a host of scientific, educational, and cultural causes, among them, energetics, monism, and the creation of a new world language (Ido).[37]

Since Arrhenius's support was decisive for the decision to award the prize to Ostwald, it is interesting to examine the reasons why he did not act until 1909. It was certainly not a fortuitous circumstance that Arrhenius threw his support behind Ostwald's candidacy only after the latter had adopted atomism. In doing so, Ostwald renounced the doctrine of energetics that he had advocated since the early 1890s, a doctrine in which energy was considered "the unique real entity in the world, and matter . . . a bearer rather than a manifestation of the former."[38] During the most active phase of the debates between atomists and energeticists in the 1890s, Arrhenius, perhaps out of regard for his old friend and teacher, did not openly attack Ostwald's energetics in writing. However, in reporting to his Swedish colleagues on the defense of energetics by Ostwald and others at the meeting of the German Society of Scientists and Physicians in 1895, Arrhenius adamantly rejected it as "speculations in *Naturphilosophie*."[39] One can understand, then, the "great joy" that Arrhenius took in participating in Ostwald's conversion to atomist views in the summer of 1908 and in reporting this event to the Swedish Society of Chemists later in the year. Arrhenius's pleasure was heightened by the fact that Ostwald resumed his belief in molecular reality after having been confronted with firm experimental evidence (of Brownian movement) for this viewpoint.[40] That Arrhenius nevertheless did not move on Ostwald's candidacy in 1908 was due (as discussed in the next section) to the priority he gave that year to having the physics and chemistry prizes awarded to Planck and Rutherford, respectively, whose work directly bore out the atomic hypothesis.

In 1909 the chemistry committee was in considerable disarray over the candidate or candidates it should recommend to the Academy. The consensus that had reigned among the nominators in the early years had dissolved, and there was now a host of candidates, none of whom had received more than a handful of nominations. For that year's prize, Ostwald was nominated by Arrhenius, van't Hoff, and G. Bredig, professor of physical chemistry in Heidelberg and a former assistant to Ostwald.[41] Shortly after the end of the nominating period, Arrhenius wrote to Widman suggesting that a prize for Ostwald would be analogous to the one given to von Baeyer in 1905 since "just as the organic chemists at that time wanted to give recognition to their great master and leader, despite the fact that his best works were somewhat removed in time, now the physical and inorganic chemists want to do the same." Having suggested to Widman the wording of the award citation for Ostwald, Arrhenius ended his letter with an argument that carried particular weight for him: In contrast to many other candidates, he stated, Ostwald was not independently wealthy and could well use the Nobel prize money.[42]

Widman must have taken Arrhenius's arguments to heart, for shortly after the spring meeting of the committee he wrote back with an encouraging report that also provides rare insight into the workings of the committee. After the other members had talked for a couple of hours, "murdering each others' candidates," he had thrown in the names of Nernst and Ostwald. Having no problem in eliminating Nernst's candidacy, he then argued for an award to Ostwald on the basis of the latter's investigation of catalytic processes. Although these had figured in one of the nominations for Ostwald in 1904, as well as in the above-mentioned proposal by Bredig, "this was a view," to quote Widman, "which the other members of the committee had not considered until now, but it evidently impressed them." The chances of success were good, but it was necessary to give particular care to the special report on Ostwald, which Widman had agreed to write only because Arrhenius had promised to help.[43] When the report came due, however, Widman found out to "his surprise and 'horror' " that Arrhenius was off in Manchester doing radioactivity research in Rutherford's laboratory. To Widman this was not a sufficient excuse for delay since, he wrote Arrhenius, "you are probably not doing anything in particular over there and since it is important that you do not bother him [Rutherford] in *his* work, you might as well write the report." If Arrhenius did not comply, Widman cautioned, he would give up Ostwald and instead write a report on V. Grignard, the French organic chemist and future Nobel prizewinner (1912) with whose work he was intimately familiar.[44]

The report on Ostwald that Arrhenius wrote after his return from England, and that Widman handed in under his own name at the fall meeting of the committee, focused on catalytic phenomena, which were described as the thread running through Ostwald's experimental work. The report shows the utility of this device, for it permitted the committee to hold up Ostwald's work on a phenomenon, or rather a conceptualization of chemical processes, that in itself was specific enough, but that also, through its applications to different branches of chemistry, was sufficiently general to subsume a significant body of Ostwald's writings. So it was that, in addition to Ostwald's work focusing directly on catalysis (*Ueber Katalyse*, 1901, which also had the advantage of being relatively recent), Arrhenius could point to his own and Ostwald's investigations on the conductivity of acids in solutions as another example of catalysis studies. Likewise, the rapidly developing field of enzyme research could be brought in under the catalytic "umbrella," making Ostwald's award a logical successor to Buchner's in 1907 for the first experimental demonstration of the role of enzymes in cell-free alcoholic fermentation. Hence, in its report to the Academy recommending Ostwald for the prize, the committee could explain its sudden recognition of Ostwald's catalysis research by referring to "the development of science proper, especially investigations into the catalytic action of enzymes and colloids, which have highlighted the importance of a group of works by Ostwald, that is, those dealing with catalytic reactions."[45] Although the arguments presented were sufficiently persuasive for Ostwald to be awarded the prize, Partington, in his *History of Chemistry*, states that "since Ostwald had no theory of catalysis, he proposed superficial analogies"; and Hiebert and Körber note that on several basic questions concerning catalytic reactions "no convincing answers were supplied by Ostwald and his collaborators."[46]

The above account shows that Arrhenius played a key role in the rewarding of the original triumvirate of Ionists. He also used his influence to delay until 1921 the awarding of the prize to Nernst who, as noted in Chapter 4, was the major problem candidacy in chemistry during the period studied here. The road of Nernst's candidacy from the first time it was proposed in 1906 until he received the prize in 1921 was the longest of any during the period for which the records are available, longer and more complex even than that of Planck, since Nernst was nominated for a greater number of years (fifteen as compared with twelve for Planck) and also received nominations in physics as well as chemistry. His candidacy will only be treated in barest outline, however, since chronologically the award falls outside the period studied here. More important, the work in chemical thermodynamics (Nernst's heat theorem or the third law of thermodynamics) for which he finally received the prize rested on quantum statistical considerations; thus its place is on the far side of the watershed represented by the award of the 1918 physics prize to Planck for his "discovery of energy quanta."

Nernst (who was closer in age to Arrhenius than to van't Hoff or Ostwald) had made a significant contribution to electrochemical solution theory in the 1880s and 1890s, notably through his work on the fundamental relationship between electromotive force and ionic concentration, which led him to resolve the age-old problem of the source of electricity in the galvanic cell.[47] During the *Sturm und Drang* period of the physical theory of solutions, Nernst and Arrhenius, who had met while they were both working under Kohlrausch in Würzburg in 1886–1887, were in close contact. At the turn of the century, however, their relations were broken off and were not resumed until the summer of 1921.[48] Arrhenius's biographers have referred both to the debate between Nernst and Arrhenius in the *Zeitschrift für physikalische Chemie* following the attempt of the latter to extend his theory from weak to strong electrolytes, and to Nernst's attack on Arrhenius's work in immunology as examples of the tensions that existed between the two.[49] These conflicts may merely have added fuel to the fire, however, for the Nernst–Arrhenius relationship was probably fraught with antagonism from the start. The two men's egos were as large as their achievements, and it seems inevitable that professional jealousies should have arisen. Nernst's work on strong electrolytes highlighted the fact that Arrhenius's theory was restricted to weak electrolytes. Nernst's successes in exploring new facets of chemical equilibrium processes from the point of view of thermodynamics (incorporated into successive editions of his popular textbook *Theoretische Chemie*) may also have piqued Arrhenius, since it showed Nernst to be the superior physicist and theoretician.[50] Moreover, Arrhenius, always the puritan in financial matters, took offense at Nernst selling the patent for the lamp (using a solid electrolyte) he had developed in the 1890s to the German firm Allgemeine Elektrizitäts Gesellschaft (AEG) for a million marks.[51]

Nernst was regularly proposed for the chemistry prize and, from 1911 onward, the physics prize as well by leading German chemists and physical chemists, notably absent among whom, however, were van't Hoff and Ostwald. Cited in the nominations were both his earlier works concerning the electromotive activity of ions in the galvanic cell and his heat theorem, formulated in 1906. Whether it was the chemistry or the physics committee that undertook the evaluation of his work, however, it would invariably be carried out by Arrhenius and, predictably, lead to a negative recommendation. Most of these evaluations were carried out in chemistry where the committee requested that they be assigned to Arrhenius in his capacity of director of the Nobel Institute for Physical Chemistry.[52] Arrhenius's close relationship with members of the committee, Pettersson and Widman in particular, undoubtedly helped, not only because they could be counted on (as stated in the letter from Widman to Arrhenius cited in connection with the Ostwald decision) "to eliminate Nernst" from the discussion, but also because Arrhenius, as part of his side of the "bargain," proposed other candidates who would be acceptable to the committee majority (e.g., Ostwald, Rutherford, and Marie Curie).

Arrhenius's support of atomic theory

The advances in atomic theory that Arrhenius promoted for the prizes were primarily those brought about by the men who founded atomic physics in Great Britain, especially J. J. Thomson and Rutherford.[53] Among Arrhenius's several interventions on behalf of the atomic theory, the actions he undertook in 1908 in support of Rutherford and (indirectly) Planck are of most interest since they show how he sought to use the Nobel prizes to highlight work that directly bore out the atomic hypothesis and illustrate the obstacles he encountered in moving the prize decisions somewhat closer to contemporary debates in the international scientific community.

In 1908 the attention of Arrhenius and many of his colleagues in other countries was focused on the rapidly accumulating evidence for the "essentially atomic" theory of matter through the determination of Avogadro's number (*N*) by different experimental methods.[54] In the summer of 1908 Rutherford and H. Geiger announced their determination of the basic unit of electric charge (*e*) of alpha particles, each particle having been found to carry two such units. This provided more precise values than those established by J. J. Thomson for the fundamental electric charge on the electron, the "corpuscle" that he had discovered in 1897. In the course of the year, J. Perrin published four papers on Brownian movement in the *Comptes rendus* of the French Academy of Sciences, the last of which contained the most accurate value to date for *N*. More important, the values obtained by Rutherford and Geiger, as well as by Perrin, agreed well with those that Planck had deduced from his theoretical formula for heat radiation first presented in 1900.[55] These developments, which (as already noted) led Ostwald to abandon his antiatomist viewpoint, were of course reported and commented on at scientific meetings and in articles in professional journals.[56] It is not surprising, then, that Arrhenius sought to crown this "year of the atom" by having the Nobel prizes in physics and chemistry simultaneously awarded to Rutherford and Planck.

In 1908, Arrhenius nominated Rutherford for the prizes in both physics and chemistry for his investigations in radioactivity, mentioning in particular Rutherford's suggestion that the alpha particle is an atom of helium carrying a double positive charge. In physics, there were four other nominations of Rutherford, including ones from Planck and Warburg. In chemistry, there were two nominations of Rutherford in addition to that by Arrhenius, one of which proposed that the prize be divided between Rutherford and Soddy.[57] How did it come about, then, that despite the opinion of such influential nominators as Planck and Warburg, Rutherford received the prize in *chemistry* for his disintegration theory, construed as "predominantly physical" research?[58] This decision has puzzled observers, one of whom has stated that it must have materialized "through some misunderstanding of the nature of his achievement."[59] Before considering the complex sequence of events that led to this decision, it is of interest to examine two precedents in the decision making of the committees indicating why, from the outset, the question of whether Rutherford should receive the prize in physics or chemistry was an open one.

The first precedent was the decision to award the 1903 prize in physics to Becquerel and the two Curies. As discussed earlier, the inclusion of the Curies had led to the first jurisdictional dispute between the two committees. Although this particular dispute had been resolved by deleting the discovery of radium from the award citation, the chemists had made it clear that in the future they would assert their right to make awards in the area of radioactivity. Another previous decision of some relevance was the awarding of the 1906 physics prize to J. J. Thomson. The rewarding of Thomson clearly had had to precede that of Rutherford not only because the former was senior in age and scientific standing, but also because of Thomson's pioneering work in atomic physics. This decision was significant for, as shown in Arrhenius's lengthy report on Thomson's work, his arguments in favor of atomism could, when couched in the right language, be made acceptable to the members of the physics committee. It is noteworthy, though, that whereas the main part of Arrhenius's evaluation concerned electron theory, the award citation made no reference to Thomson's "corpuscles" but mentioned only his "theoretical and experimental investigations on the conduction of electricity by gases."[60]

The deliberations of the chemistry committee in the spring of 1908 show that the chemists moved with some dispatch to ensure that Rutherford (alone or together with Soddy) received their prize. At its last meeting before the summer recess, the committee decided to commission a report on the two to be used as the basis for a prize recommendation.[61] This was usually an indication that a consensus was emerging, the report being less an evaluation than an argument for the candidate(s) in question. The task of writing the report was assigned to Söderbaum. A close analysis of the report's content, language, and style of argument suggests that it was written in whole or in part by Arrhenius and handed in under Söderbaum's name, just as Widman was to hand in Arrhenius's report on Ostwald in the following year. Arrhenius's involvement with the report is apparent in several respects, among which are the following.

First, the report shows an intimate knowledge of recent developments, both theoretical and experimental, in the rapidly moving field of radioactivity, and especially, a grasp of what the Rutherford–Soddy disintegration theory had contributed to the understanding of atomic structure, neither of which Söderbaum, an agricultural chemist, was likely to have possessed.

Second, the report shows the clarity of argument and presentation of technical data characteristic of Arrhenius's style. This was a matter of some importance when it came to convincing the uninitiated members of the Academy.

Third, and most important, the report does not argue strongly that Rutherford's work was all that significant for chemistry. It is twice stated, for instance, that the experimental part of Rutherford's work belonged to the area of physics. As for the theoretical part, the disintegration theory was presented as building on the work of atomic physicists – J. J. Thomson, J. Larmor, and H. A. Lorentz – and as contributing importantly to an understanding of atomic structure. Only at the end of the report does its author refer to the significance of Rutherford's work for chemistry: The disintegration theory, it is stated, has radically changed the foundation of chemistry because the tenet of the immutability of the elements no longer holds and because the inner structure of the atom now represents a problem that can be made the object of a "completely realistic, scientific discussion carried out on the basis of exact measurement."[62]

To understand why Arrhenius may initially have wanted to leave the door open for rewarding Rutherford in physics, but did not do so, it is necessary to turn to the deliberations of the physics committee.

Here, the attention was focused on two alternative sets of candidates: W. Wien and Planck, who had been nominated for "their work in the theory of temperature radiation"[63] by I. Fredholm, professor of mechanics and mathematical physics at the Högskola; and G. Lippmann, proposed for his method of photographically reproducing colors, in a nomination from the French Academy of Sciences. Fredholm could hardly have given his proposal much chance of success, for the research area he introduced in his one-paragraph letter was one of the more esoteric ones in theoretical physics, and one with which no member of the physics committee was intimately familiar. The research question at the source of Wien's and Planck's theories, known as the black-body problem, consisted of determining the intensity of heat radiation at different wavelengths in a cavity with perfectly absorbing (i.e., black) walls maintained at a fixed temperature. While Planck attacked the problem from a mathematical – theoretical viewpoint, invoking principles of thermodynamics and electromagnetism, Wien combined theoretical work with experiments. At the same time, the precise measurements that made empirical verification of the hypotheses possible were carried out by experimental physicists (O. Lummer and E. Pringsheim). It was the theoretical work, though, that would bear the most fruit. Planck's formula implied a radical new conception of the emission and absorption of energy, which seemed to take place not, as hitherto assumed, in a continuous manner, but instead discontinuously, in discrete packets or "quanta."[64]

When the Fredholm proposal came before the physics committee in the spring of 1908, Arrhenius took on the task of writing the special report on Wien and Planck. The main thrust of the report was to argue for rewarding Planck alone by breaking the connection between Wien's and Planck's work. Arrhenius made what he considered an important distinction between Planck's derivation of a new radiation formula, which he considered a correction of Wien's radiation law, and Planck's use of the constant k (Boltzmann's constant) in the new radiation formula to calculate the atomic constants e and N. He linked this to a discussion of the similar values that Planck and Rutherford had obtained for these and went on to state, "In this way . . . it has been made extremely plausible that the view that matter consists of molecules and atoms is essentially correct, in spite of recent objections by, e.g., [F.] Wald and Ostwald. No doubt this is the most important offspring of Planck's magnificent work."[65]

On the basis of Arrhenius's report, the committee decided that if the prize should be given for work in heat radiation, Planck alone should receive it. The "if" is important, for this recommendation was accompanied by a statement from the committee majority (i.e., all the members except Arrhenius) expressing their doubts as to the wisdom of rewarding only the theoretical side of work on heat radiation rather than dividing the prize between "the foremost theoretician and the foremost experimentalist, in this case probably Lummer." Since this was not possible in view of the fact that O. Lummer had not been proposed for the prize, a more prudent course of action, they suggested, might be to postpone the decision on Planck, especially since Lippmann of France was a good alternative candidate.[66]

To sum up, it seems likely that Arrhenius seized the opportunity to reward Planck alone for works that supported the atomic hypothesis. He therefore did not press Rutherford on the physicists but left him to the chemists who, in any case, would be more accommodating to the candidacy of one who represented the atomist viewpoint and who was moreover the leading figure in radioactivity research.

Initially, Arrhenius's strategy seemed to be paying off, for despite the caveat that the physics committee had attached to its recommendation for Planck, it was endorsed by the physics section of the Academy with all members except one voting in favor. When the Academy met in plenary session, however, it decided by a substantial majority to award the prize to Lippmann rather than to Planck.[67] Since Nagel has given a full account of the discussion in the Academy,[68] only the major considerations will be treated here.

The circumstances that deprived Planck of the 1908 prize were linked to his hypothesis of the elementary quantum of energy. Arrhenius's report contained no reference to energy quanta (or an equivalent term). It seems unlikely that Arrhenius was ignorant of this aspect of Planck's radiation theory, but he probably chose to forego mentioning it as too speculative, and in any event his attention was focused not on Planck's theory but on its use to calculate the atomic constants e and N.[69] When the physics committee recommendation was discussed in the physics section of the Academy, Carlheim-Gyllensköld made reference to Planck's "new, previously unimagined idea ... of the atomistic structure of energy,"[70] but this was probably not sufficiently specific to cause any hesitancy about Planck's hypothesis. The situation was different in the Academy meeting in plenary session to reach a final decision, for present here were some mathematicians led by Mittag-Leffler, who not only had a better grasp of mathematical physics but had attended the International Congress of Mathematicians in Rome in the spring of 1908 and probably heard Lorentz's lecture on the current state of radiation theory.[71] Mittag-Leffler used the uncertainties surrounding the theory of quanta as the chief argument against awarding the prize to Planck alone. "Planck's derivation" of his radiation law, he said, "is based on a totally new hypothesis which can hardly be considered plausible, namely that of the elementary quantum of energy.... an evaluation of its worth at present raises great difficulties.... the deferment of a definitive judgment is therefore to be preferred."[72] Although Mittag-Leffler, according to his diary, had set out to restore the original Fredholm proposal for dividing the prize between Wien and Planck, his attack on Planck was probably the main reason why the Academy chose Lippmann as a compromise prizewinner. Mittag-Leffler was probably not averse to this solution, and in any case he had the satisfaction, also noted in his diary, of seeing Arrhenius go down to defeat.[73]

At the opening of the debates in the Academy, Arrhenius had arranged things so that, exceptionally, the order of the prize deliberations was reversed and the prize in chemistry discussed before that in physics. Although Arrhenius did not give any reason for his request, he may have used this tactic in order that Rutherford would carry Planck to the prize on the basis of the links Arrhenius had tried to establish between their work in support of the atomic hypothesis. This explanation seems plausible in view of the fact that Rutherford's candidacy in chemistry rested on firmer ground than that of Planck in physics. After Arrhenius had reported on Rutherford for the chemistry committee, that body had agreed to recommend Rutherford for an award. The proposed award citation stressed the importance of his disintegration theory for the chemistry of radioactive substances. The chemistry section of the Academy had endorsed this recommendation, and it also became the final decision of the Academy. That there were some lingering doubts on the wisdom of this choice, however, is illustrated by the comment of one of the members of the physics section who felt that "the physicists had been tricked into handing over their best candidate, Rutherford, so to speak, against his own wishes, to the chemists."[74]

There is a final ironic twist to the story of the passing over of Planck for the 1908 physics prize that suggests that Planck may have been privy to Arrhenius's scheme to use the prizes to promote work confirming the atomic hypothesis. Following the meeting of the physics section of the Academy, reports were widely circulated both in the popular press and in the journal *Nature* that Planck was the sure winner of the year's physics prize. In an interview Planck gave to the *Leipzig Neueste Nachrichten*, he stated that he did not yet consider himself a prizewinner, but that if the press reports were true,

I presume that I owe this honor principally to my works in the area of heat radiation. For some time I have been occupied with establishing the absolute weight of the atom and I have now succeeded in arriving at a positive result. One assumed earlier that the absolute weight of the atom could not be established. We are now out of the area of speculation however and have reached firm ground.... What is interesting here is that Rutherford, who is primarily concerned with research on radioactive bodies and substances, through quite different means than I myself has arrived at approximately the same value.[75]

There is no doubt that the entrance of the Nobel Committee for Physics and the Academy into the field of quantum theory (premature, but not much so, for in 1909 Lorentz himself had taken up energy quanta[76]) had the effect of hardening positions on the question of Planck's award, a matter that was to occupy the committee until 1919.[77] During the remainder of the period up to the First World War, Arrhenius did not again give his support to Planck's candidacy. This may have been because, in 1910, Nernst's heat theorem was linked to and shown to be in agreement with predictions from quantum theory, specifically Einstein's formula for the specific heat of solids.[78] In the context of decision making in the physics committee, it meant that the uncertainties of most of the experimentalists on the committee about Planck's theory could be exploited by Arrhenius in arguing against Nernst, who was now regularly proposed for the physics prize, in 1911 by Planck himself.[79]

During this period, Arrhenius's actions were more often directed at the chemistry committee, where he continued to support candidates whose work represented advances in atomic theory. Two of his more important interventions deserve brief mention. In 1913 he gave his support to A. Werner, the Swiss chemist whose theory of the bonding of atoms in so-called complex compounds represented an immense effort to systematize the chemistry of inorganic complexes. Here, Arrhenius's authority undoubtedly helped members of the chemistry committee overcome their hesitation over rewarding a theory which was neither complete nor entirely verified by experiment. However, as Arrhenius stated in his nomination of Werner, if "we are to demand perfection in all details of work proposed for the prize, we should in all probability be forced to turn down everything."[80] Shortly after Werner was awarded the chemistry prize of 1913, his theory was tested experimentally by the powerful new method of X-ray crystallography, an important early application of which turned out to lie in the area of atomic structure. In the physics committee, attention was focused in 1914 and 1915 on the three men (von Laue, W. H. Bragg, and W. L. Bragg) who had contributed most significantly to the discovery of the diffraction of X-rays by crystals. For the 1915 prize in chemistry, Arrhenius proposed H. G. J. Moseley, a member, as he put it, of "the flourishing Manchester school" whose study of the x-ray spectra of the elements shed entirely new light on the periodic table and was also important in the elaboration of Niels Bohr's atomic model.[81] The report on Moseley that Arrhenius wrote as an adjunct member of the chemistry committee contained a detailed analysis of recent advances in atomic theory. Here, for the first time, but in a report for the chemistry and not the physics committee, Rutherford's and Bohr's atomic models were discussed with reference to spectroscopic data provided by Rydberg and Moseley. Although the chemistry committee took a positive view of Moseley's work, which "undisputably will be of utmost importance for theoretical chemistry," it felt that in view of its recent date, his award could wait.[82] Nobody could have known, of course, that in the same year (August 1915), Moseley would die in the Battle of Gallipoli.

THE CAMPAIGNS OF MITTAG-LEFFLER

Opening the doors to theory

From the beginning, the campaigns of Mittag-Leffler were undertaken with a twofold purpose: to promote works in mathematical and theoretical physics and, by doing so, to ensure that Poincaré would one day receive the prize. Since he also had a strong interest in helping his fellow members of the French Academy of Sciences win prizes, it is not surprising that he was sometimes working at cross-purposes. In 1902, however, this was not yet so.

The campaign that Mittag-Leffler undertook in that year concerned H. A. Lorentz, whose chair in theoretical physics at the University of Leyden was one of the first created in that specialty and whose reputation rested on his electron theory, elaborated in the 1890s and widely adopted around 1900.[83] Mittag-Leffler's aims were clearly brought out in the letter he wrote to Painlevé once the campaign was under way: "If I succeed . . . I will have *opened the doors to theory* . . . and then first Poincaré and later you yourself will follow." In part, the campaign was also undertaken to jeopardize Arrhenius's candidacy in physics, which Mittag-Leffler described as a ridiculous idea. In assessing his chances of having the prize awarded to Lorentz, he cites the physics committee as the main obstacle. It was "composed of nonentities [*nullités*] with the exception of Arrhenius who understands nothing about physics and who wants the prize for himself. . . . Just think," he wrote, "the young Ångström who is an able experimentator but who does not have a thought in his head"; Hildebrandsson, who was "very active in meteorology but with no patience for physics"; Hasselberg, "an excellent friend" who "only knows spectroscopy." "You have to admit," he wrote, "that I shall have every reason to be proud if I manage to get them to propose Lorentz."[84]

The idea of proposing Lorentz for the prize came to Mittag-Leffler from Poincaré, who suggested it when the two met in Paris in November 1901. Once he had returned to Stockholm, Mittag-Leffler began writing Poincaré giving him his "instructions." To ensure success, Poincaré should write a detailed report citing one or two remarkable works by Lorentz. This was indispensable "because one would hesitate to reward his entire *oeuvre*." The report should also mention Zeeman because "there are those who push for Zeeman." In the references to Zeeman's work, Poincaré should claim that it was "undertaken at the instigation of Lorentz to exemplify his theories" and "that Zeeman was a student of Lorentz, etc." Once he had written his report, Poincaré should have it signed by as many physicists and mathematical physicists in Paris as possible. He should subsequently hand it over to Mittag-Leffler, who would collect signatures in Germany and England. Röntgen had already been contacted and "would be happy to join us."[85]

The short report that Poincaré wrote supporting his proposal of Lorentz hardly met the requirements of the statutes, for it neither mentioned any specific "discovery or improvement" nor did it make a reference to any of Lorentz's published works. That Poincaré should make Lorentz's electron theory the cornerstone of his proposal is hardly surprising, though, for he was intimately familiar with Lorentz's work in the area of electromagnetism and optics, which he had made the subject of his lectures at the Sorbonne in 1899. Repeating his views – most succinctly presented in his address to the International Congress of Physics in Paris in 1900 – that theories are not true or false but only more or less useful,[86] he made only a short reference to the particulate concept of electricity that lay at the basis of Lorentz's electron theory.[87] "We know nothing about what this hypothesis is worth in itself," he wrote, "and we cannot do anything about this. What we have to ask ourselves is whether or not this hypothesis has been productive and useful in bringing together existing facts or in predicting new ones." On this latter score, he pointed to two areas where he felt that Lorentz's work had filled gaps left by earlier theories, in particular those of Maxwell and Hertz. One concerned the inability of these theories to give a satisfactory explanation of problems associated with the movement of the earth through the ether, in particular the phenomena of aberration and ether drag. These had been resolved, Poincaré stated, by Lorentz's assumption of a stationary ether, in which the propagation of light is due to moving electrons, as well as by his concept of "local time."

The other theoretical gap lay in explanations provided for magnetooptic effects, the first such effect having been discovered by Faraday in 1845. While Faraday had demonstrated the effect of magnetism on propagated light, Zeeman was the first to show, in 1896, the influence of magnetism on the emission of light by the broadening of the spectral lines of a sodium flame burning in a magnetic field. The Zeeman effect, as it came to be known, was explained by Lorentz's electron theory, which was also used to work out refinements in Zeeman's discovery.

Having stated this, Poincaré immediately had to concede that subsequent experiments, rather than providing further confirmation of Lorentz's theory, "had led to its ruin." Following Zeeman's lead, between 1897 and 1900 other spectroscopists – T. Preston in England, A. Cornu in France, and C. Runge and F. Paschen in Germany – had found that many spectral lines in a strong magnetic field did not form the triplet of distinct polarized components predicted by Lorentz's theory, but resolved into quadruplets, sextuplets, or more. In 1899 Lorentz reformulated his theory to account for the quadruplet, but the problem of further complexities was not to be resolved entirely until the advent of quantum theory.[88] In his lecture courses of the same year, Poincaré stated that Lorentz's reformulated theory could, *à la rigeur*, account for the quadruplet but could go no further.[89] After having mentioned the "ruin" of Lorentz's theory in his proposal, he tried to paper over these difficulties by arguing that the complexities noted above, although not predicted by the theory, could nevertheless be accounted for by some slight alterations of the original hypothesis. Since to Poincaré's mind, theories were inherently fragile and since Lorentz's theory had served its purpose by predicting the Zeeman effect, he saw no need to defend it further but ended his proposal instead on a general note of skepticism that deserves to be quoted in full:

Theories are often accused of fragility, and, doubtless, if they claimed to reveal to us the essence of things, the spectacle of so many ruined theories would be enough to turn us skeptical. But when Lorentz's theory has been entombed with those of his predecessors in this great graveyard, will not the facts which he foresaw and which his theory caused to be discovered continue to exist? And if one day his theory should be abandoned, how wrong will be those who say that if his theory envisaged true facts this was merely by chance since the theory itself was false. No, it is not by chance, it is because it has revealed to us hitherto unknown relationships between facts seemingly foreign to one another, and because these relationships are real, and will continue to be real even if electrons should no longer exist. Such are the truths one may hope to find in a theory, and these truths will live on after him. It is because we believe that Lorentz's works contain many truths of this nature that we propose they be given their due recompense.[90]

By itself, Poincaré's proposal was hardly designed to convince the committee and bring about the "victory of theory" desired by Mittag-Leffler, so much would depend on how the campaign was conducted. Following instructions, Poincaré had the proposal signed by several members of the French Academy of Sciences, only three of whom, however, were official nominators: A. Cornu, G. Lippmann, and J. Boussinesq. Having received the proposal, Mittag-Leffler affixed his own signature and then added the names of other scientists – among them Röntgen and Planck – who had made it known by cable or letters that they were in agreement. Lorentz thus received a total of ten nominations, all prompted by Mittag-Leffler's campaign. However, at its August meeting the physics committee declared the cables received in support of the proposal invalid, thus reducing the nominations to six, a number that still made Lorentz the most nominated candidate that year.[91]

Success would come, though, not so much through the efforts

of Mittag-Leffler but, as he had suspected from the beginning, through the linking of Lorentz's name to that of Zeeman. Zeeman had not been officially proposed for the prize in the campaign, since in Mittag-Leffler's opinion "Zeeman could have received the prize another year."[92] Although Zeeman had received two nominations in 1901, when the nominating period for the physics prize of 1902 was drawing to a close there was no formal proposal of Zeeman at hand. It is curious that Zeeman should not have been proposed by any of the spectroscopists on the committee, since, to quote Lorentz's Nobel lecture, his work and theirs together helped illuminate "how wonderful and rich a world those investigations into spectra has opened up to us."[93] Instead, it was Arrhenius who, at the last minute, handed in the one nomination of Zeeman that made it possible to consider the latter for the prize.[94]

That the committee's recommendation of Lorentz depended largely on including Zeeman in the prize is brought out in the short report that Ångström wrote to the academy justifying the proposal. Here the Zeeman effect, hailed as being "among the discoveries that have brought about the most significant advance of physical science," was given extensive treatment. Lorentz's electron theory was described much more summarily as "a simple hypothesis" that traced its origins to the fundamental equations of Maxwell and whose chief merit was that it explained the Zeeman effect. Following Poincaré's lead, the complexities revealed by subsequent investigations were given short shrift, since "these have distracted neither from the discovery nor from the merit of the theory which had the power of predicting it."[95] The bias in favor of Zeeman, the experimentalist, as against Lorentz, the theoretician, reflected in Ångström's report, was shared by the majority of the members of the committee. Coupling Lorentz's name with that of Zeeman was an important condition for the awarding of the prize to the former, chiefly because of the committee's experimentalist bias but also because the Zeeman effect could be held up as a "discovery" that satisfied the conditions of the statutes. (See the section, "The use of statutory criteria in committee decision making" in Part I of Chapter 6.)

The battle over Poincaré

Mittag-Leffler did not directly proceed to set up the nominations for Poincaré, despite what he termed the "halfway satisfactory" result of the campaign for Lorentz, since it was now the turn of an experimentalist.[96] In 1903, not one but three experimentalists – Becquerel and the Curies – were awarded the physics prize. Since they were all French, some time would have to elapse before Poincaré could be put forward for the prize.

Mittag-Leffler was all set to go to battle in the Academy for Becquerel and the Curies and against Arrhenius, but as he noted in his diary, this was unnecessary, for Ångström "had managed the campaign with great proficiency."[97] There is no doubt, though, that Mittag-Leffler played a role in assuring that Marie Curie was one of the three prizewinners. The initial position of the committee on this matter is not known, but it is certain that if it had relied on the nominations, her inclusion would have been uncertain. In 1903, only the names of Becquerel and Pierre Curie had been put forward in a proposal that clearly emanated from the French Academy of Sciences, since it was signed by Darboux and Poincaré together with several members of the academy who had not been invited to nominate.[98] Marie Curie, of course, was not (and would never become) a member of the academy.[99] The possibility that Mme. Curie might not share the prize with her husband must have appeared sufficiently great for Mittag-Leffler to write to Pierre Curie. The latter replied: "If it is true that one is seriously thinking about me [for the prize], I very much wish to be considered together with Madame Curie with respect to our research on radioactive bodies." He pointed to the important role she had played in the discovery of the new elements radium and polonium and added: "Don't you think that it would be more artistically satisfying [*plus joli d'un point de vue artistique*] if we were to be associated in this manner?"[100] As it turned out, the committee recommended that the Curies together receive half of the physics prize. Since Marie Curie had not been nominated in 1903, however, the committee had to go back to the nominations of 1902 and resurrect one made in favor of both Curies by C. Bouchard, professor of pathology at the Paris Faculty of Medicine. Although not strictly in accordance with the statutes, this procedure was considered defensible since Bouchard, as a foreign member of the Royal Academy of Sciences could be held to have permanent nominating rights.[101]

It was not until 1909 that Mittag-Leffler felt the time was ripe to start a serious campaign for Poincaré. This delay and the strange manner in which the campaign was launched can be explained by the unexpected decision to award the 1908 physics prize to another French candidate (Lippmann) instead of Planck.

In December 1908, Mittag-Leffler wrote to Painlevé, (professor of mathematics at the Ecole Polytechnique), who had become his main correspondent in France: "One can not propose Poincaré for the prize immediately after Lippman, but I think such a proposal would be successful in 1910 if we prepare it well ahead of time." In the meantime, he presented Painlevé with another scheme that is interesting chiefly because it shows the roundabout way in which Mittag-Leffler went about trying to secure the prize for Poincaré. In the letter quoted above, he wrote: "Do you think one could give the Nobel prize for the invention of the airplane? In that case, to whom should one give the prize? To Wright, to Farman, or to both of them?"[102] Painlevé, who had been closely involved with the several important events in aviation that had taken place in France in 1908, answered enthusiastically with a detailed analysis of the current state of the art of aviation and of the credit due to the different pioneers.[103] He concluded: "After pondering the matter, I think that the prize should be divided in two equal parts: one going to the Wright brothers, and the other to the Voisin brothers [airplane constructors] and to Farman, the pilot of the Voisin machine." Poincaré was in total agreement with this plan, Painlevé wrote, and would sign the detailed proposal that Painlevé was drafting and would shortly submit to Mittag-Leffler.[104]

At year end, when Painlevé's proposal was ready to be signed, trouble arose. When he presented the proposal to Darboux and Lippmann for their signatures, Painlevé discovered to his great consternation that they were putting forward Poincaré for the prize. Darboux took out of his briefcase "a report on Poincaré as a physicist, already covered with signatures, which had been prepared three weeks ago." Why had he not been consulted, Painlevé asked Darboux, who answered rather sheepishly that he did not consider it wise to have the nomination supported by mathematicians. Painlevé also found it extraordinary that Poincaré himself had not informed him of Darboux's proposal, especially since the aviation proposal could be interpreted as a scheme to deprive Poincaré of the prize. In the end, only he and Poincaré signed the aviation proposal, which he sent on to Mittag-Leffler to use as he saw fit.[105]

In answering Painlevé's letter, Mittag-Leffler not only came up with a plausible explanation for Darboux's action but also proposed a scheme that would turn a touchy situation into one that in the end would enhance Poincaré's chances of obtaining the prize. Mittag-Leffler surmised that Arrhenius had promised Lippmann, when the two met in Stockholm at the Nobel prize ceremonies, that he would support Poincaré for the prize and had advised Lippmann that since success depended on Poincaré being supported by physicists, it would be better if the mathematicians were kept out of the matter. Arrhenius was right to suggest this approach, Mittag-Leffler felt, but manifestly Arrhenius's promises could not be counted on; this year, for instance, he had pushed a German (Planck) rather than a Frenchman (Lippmann). The counterplot proposed by Mittag-Leffler was the following: Painlevé should maintain his nomination of the aviators, and he (Mittag-Leffler) would sign it and make sure that it was signed by other Swedish mathematicians. This would convince the experimentalists among the physicists in the Academy "that I am against Poincaré and that I am wildly enthusiastic about aerial navigation." Only when this fiction was well established could he launch a real effort on behalf of Poincaré.[106]

Proceeding according to plans, the aviation proposal or "heavier than air," as it was called, was submitted in the names of Mittag-Leffler, Poincaré, Painlevé, and four mathematicians at the Högskola.[107] In the physics committee, however, the proposal met with skepticism. Pointing to the loss of life that would ensue even from minor errors in the construction or navigation of airplanes, the general report of the committee (written by Arrhenius) concluded that "in its present state, this invention can hardly be considered to be of benefit to mankind."[108] Notwithstanding this unfavorable opinion, Mittag-Leffler and some of the other mathematicians in the Academy voted for the proposal that only gathered a tiny minority (four votes) against the overwhelming majority of some fifty votes cast for Marconi and Braun, who received the prize in 1909. According to Mittag-Leffler, Poincaré received five votes, since all members of the mathematics section had refrained from voting for him, in order to "conceal the ruse [*masquer le jeu*]."[109]

When Mittag-Leffler launched his "real effort" to have Poincaré selected for the prize in 1910, he again carefully avoided any visible role for either himself or his fellow mathematicians in the Academy. The main proposal, developed by Fredholm in consultation with Mittag-Leffler, carried only the signatures of Fredholm, Darboux, and P. Appell, the last a professor of mathematical physics and dean of the Paris Faculty of Science. It was essential, Mittag-Leffler wrote to Appell, that no French candidate other than Poincaré be put forth and that all proposals for Poincaré should be those of official French nominators. The letter ended with a word of caution for the nominators: "Avoid as much as possible 'mathematics' and speak, instead, of 'pure theory.' Since the Nobel committee is composed of experimentalists, its members are wildly fearful of mathematics."[110]

With the Fredholm–Darboux–Appell proposal in hand, Mittag-Leffler launched what he termed his "worldwide campaign" for Poincaré. Accompanied by a form letter in which Mittag-Leffler stated "that the time has come to give the physics prize to a pure theorist," the proposal was sent to some fifty invited nominators, including all the Nobel laureates in physics, asking for their signatures.[111] In all, the campaign yielded some thirty-four nominations, the highest number for any candidate in a single year during the period 1901–1915. One-third of the nominations came from France, where both Marie Curie and Lippmann, the two surviving physics laureates, answered the appeal.[112] The University of Madrid (invited in accordance with para. 1.5 of the special regulations) produced seven chairholders in physics, all in favor of Poincaré. Mittag-Leffler's appeal also yielded two nominations of Poincaré by Italian scientists (P. Blaserna and V. Volterra). If the response from Latin countries was favorable, that from Germany and Great Britain was much less so since it resulted only in a few nominations, none of them – with the exception of Röntgen – from well-known physicists.

In Great Britain, Mittag-Leffler's campaign even aroused a certain amount of hostility. Arthur Schuster, former professor of physics at Manchester University, replied to Mittag-Leffler's appeal by declaring that he did not think "the discussions presently under way [to gain support for Poincaré] were conducive to the peaceful pursuit of science."[113] Rutherford also withheld his support, not because he had any doubt about the great value of Poincaré's contributions to mathematical physics but because he deplored what he considered to be a tendency to extend the area of physics to cover a wide range of topics, as evidenced by recent awards to "inventors and workers on technical subjects." If mathematics and mathematical physics "are to come under the scope of the physics prize," he wrote, "it will be exceedingly difficult to do justice to the claims of the physicist proper."[114] These fears were echoed by J. J. Thomson, who wrote to Arrhenius expressing concern that the broadening of the prize area would have the effect of making "the chance of getting a prize correspondingly small."[115]

The issue of whether mathematical physics should be considered "physics" was of course at the heart of the matter when it came to awarding the prize not only to Poincaré but also to other representatives of this specialty. In his 1910 campaign for Poincaré, however, this issue was not Mittag-Leffler's chief worry, for in the 1909 general report of the committee Poincaré's candidacy had been favorably mentioned – although "the committee had not seen fit . . . to place it before other ones."[116] When the Poincaré question came before the committee in 1910, Mittag-Leffler treated the favorable mention in the nature of a precedent to bolster support for his candidate. Writing to Hasselberg and Carlheim-Gyllensköld, the two main supporters of Poincaré on the committee, he stated: "The committee cannot be so inconsistent as to disown its own actions by now declaring that Poincaré's work is not physics," and he added: "If this happens, you will all get hell in the Academy."[117] This did not happen. Indeed, the 1910 general committee report, written by Arrhenius, stated: "Since, in interpreting Nobel's testament, the term physics should be used in its broadest sense, several of Poincaré's works should be taken into consideration." He mentioned specifically Poincaré's work concerning the shape of a rotating fluid mass, which he found to be of "great importance for cosmogony." There was a problem, however, in that the works cited by the majority of the nominators as most worthy of the prize were those that "concerned philosophical and mathematical questions and hence did not represent 'discoveries' or 'inventions' in physics even when these terms are used in their broadest sense."[118]

The problem of which parts of Poincaré's monumental *oeuvre* should be singled out as most worthy, and particularly which of these constituted "discoveries," had plagued the campaign from the beginning. In earlier proposals, the totality of Poincaré's publications in theoretical and mathematical physics had been submitted for appraisal; by contrast, the Fredholm–Darboux–Appell proposal of 1910 had singled out Poincaré's "discoveries concerning the differential equations of mathematical physics." The argument in the proposal went as follows: Modern physicists needed mathematics, not only to bring order into a constantly growing body of experimental data but also to formulate hypotheses that would reveal the nature of the phenomena under study. Poincaré's work on partial differential equations was intimately linked to the concerns of experimental physicists in both the methods used and the results obtained. This was shown by the fact that the starting point for his most original method of analysis – the "sweeping-out process" – was the Dirichlet problem, which consisted of finding a formula to describe the equilibrium state of electricity in a conductor. Poincaré had been the first to obtain a generalized solution to this problem. Under his influence, partial differential equations had become "a unique instrument which he has perfected ... so that it meets the ever-increasing needs of modern physics."[119] In addition to the main proposal and the many letters supporting it, a separate nomination for Poincaré by Carlheim-Gyllensköld cited three specific discoveries, among them the works in cosmogony and celestial mechanics, which figured in the general report by the committee cited above.

From a formal point of view, then, once the committee had admitted mathematical physics as part of physics, it could well have selected, as it had done in the case of Poincaré's nomination of Lorentz, one of the works cited in the nominations as the basis for a recommendation and eventual award citation. That it did not do so has to be attributed to the dissension that reigned within the committee as well as an unforeseen event – the death of Ångström – that proved to have weighty consequences for the final outcome. Ångström's pivotal role was stressed by Hasselberg, who after having assured Mittag-Leffler of his (Hasselberg's) support, stated: "Success will come if I can only convince Ångström. Then Granqvist and Hildebrandsson will join us."[120] However, as serious committee discussion about the prize of 1910 was about to begin, Ångström passed away after a brief illness. Almost immediately, the Academy proceeded to fill the vacant place on the committee. Carlheim-Gyllensköld (who had been nominated to fill the vacancy by Arrhenius) was elected, receiving fifteen votes against three cast for Fredholm, who had been put forward by

Hasselberg.[121] Although Carlheim-Gyllensköld had entered his own nomination for Poincaré and was also selected to write the special report on his works, being new on the committee he clearly did not have the ability of his predecessor to sway the Uppsala representatives.

In any event, Carlheim-Gyllensköld's efforts in support of Poincaré were doomed to fail, for at the closing of the nominating period Röntgen, probably knowing that Ångström was seriously ill, had nominated Ångström for the prize of 1910 for his work concerning solar radiation.[122] The Uppsala members of the committee seized this opportunity to pay tribute to their colleague for his work in precision measurement and probably also for the services he had rendered the committee. Since Ångström had died *after* the end of the nominating period, his case represented an exception to the prohibition of posthumous awards as specified in para. 4 of the statutes; the year 1910 was the one and only occasion, however, on which he.could be awarded the prize. Guarding against the possibility that the Academy might not want to award the prize posthumously, the committee recommended both Ångström and an alternative candidate, J. D. van der Waals, the latter for his work on the corresponding states of gases and liquids, which was hurriedly analyzed by Arrhenius.[123] A committee majority of Granqvist, Hildebrandsson, and Arrhenius supported the Ångström–van der Waals proposal, whereas Hasselberg and Carlheim-Gyllensköld together presented a dissenting opinion in favor of Poincaré. The physics section adopted the committee recommendation without pronouncing itself in favor of either Ångström or van der Waals, whereas the Academy opted for the latter. The only available information on the voting figures is Mittag-Leffler's report to Darboux that "a handsome number of votes" were cast for Poincaré in the Academy.[124]

In the 1910 general report of the committee, written by Arrhenius, the door had been left slightly ajar for a future award to Poincaré, citing his work in the area of celestial mechanics, specifically his mathematical representation of the shape of a rotating fluid mass submitted only to the forces of gravitation. The opportunity suggested by the report was seized on by Carlheim-Gyllensköld who in 1911 presented a strong argument for an award to Poincaré based on his work in cosmical physics and geodesy, the latter being of course Carlheim-Gyllensköld's own area of research interest.[125] In 1911, however, the Uppsala physicists, now joined by Hasselberg, were again active in attempting to award the prize to one of their colleagues. This time, the committee recommended its newly elected member, Allvar Gullstrand (1862–1930), professor of physiological and physical optics, citing his work in geometric optics, especially the dioptrics of the eye. Before the Academy had had time to act, however, it learned that the Karolinska Institute had decided to award Gullstrand the prize in physiology or medicine; as a result the matter was returned to the committee, which changed its recommendation to Wien.[126]

The failure to secure the prize for Poincaré, who died in 1912, marked the end of the close involvement of Mittag-Leffler with the decisions. This failure may also have made the members of his French network sense that he had overestimated his power and influence. It is significant that in 1911 it was Arrhenius who was feted in Paris: He lectured at the Sorbonne and was elected to the French Academy of Sciences. While this did not bring the change of mind with respect to Poincaré that might have been hoped for, it benefitted another French candidate, for later in that year Marie Curie was awarded her second prize, this time in chemistry. Now it was Arrhenius's name rather than Mittag-Leffler's that was linked to Darboux's in the nomination. In the remaining years preceding the outbreak of the First World War, political matters, particularly the strengthening of Swedish defense, increasingly came to occupy Mittag-Leffler's time and energies. During the war he actively supported Germany and acted as an intelligence source for the German government. In doing so, he was mainly motivated by his opinion that Finland – his wife's homeland and the country where he had resided as a young professor – could only be liberated from Russian subjugation if Germany won the war.[127] His pro-German stance could hardly have escaped the notice of his French friends, particularly Painlevé, who had served as minister of defense during the war. In 1922, when Mittag-Leffler was going through Belgium by train on his first continental journey after the war, he ran into Painlevé. At first, the latter did not recognize him.[128] Was this possibly voluntary or was it an oversight?

This chapter has described the attempts of Arrhenius and Mittag-Leffler to bring to bear what they considered were significant new developments in physics and chemistry on the prize selections and how these efforts were affected by conditions of decision making that often lay outside their control. The commanding role that the Nobel Committees for Physics and Chemistry assumed over the proceedings was the most important of these conditions, for it placed decision making in the hands of a small group without whose acquiescence such attempts as those made by Mittag-Leffler to use the nominators as a pressure group were doomed to fail. That Arrhenius enjoyed greater success in his actions, despite the fact that they received no organized support from the nominators, was mainly due to his using his influence *within* the committees.

6

Committee decision making

For the Nobel committees to be able to make annual recommendations for prizewinners clearly required the creation of procedures and criteria that would make this difficult task possible, particularly in the years when the statutory obligation that the works honored "possess the preeminent excellence . . . manifestly signified by the Will" could not be fulfilled. How the committees developed the routine decision-making practices that allowed the institution to function in a continuing and orderly manner is the subject of this chapter. The committees had to establish their operating relationship with the Royal Academy of Sciences and then function within the framework of the statutes. They also had to work out a consensual mode of decision making without which they could have neither maintained their autonomy with respect to the Academy nor fulfilled their mandate in the matter of prize selections.

Part I: Procedures and criteria for choosing the prizewinners

The specific procedures guiding prize selections grew out of a series of decisions concerning many practical problems that had not been adequately dealt with in the statutes or special regulations. These decisions were made either by the committees themselves or by the Academy in consultation with the committees. They concerned practically every aspect of the process of prize selection; their overall effect was to ensure that the committees controlled most of this process. The major factors making for such control will be discussed in the first section below.

By contrast to the procedural rules, the criteria for prize selection laid down in the statutes were never the subject of separate decisions. Instead, they formed part of the evaluations of candidates and of successive prize recommendations, each of which set a precedent for subsequent ones. Since the statutory criteria were not made explicit, it is not possible to know exactly how they were interpreted, if (which seems doubtful) they were ever stringently applied. Still, the committee reports contain enough references to these criteria to give an impression of how the work of the committees reflected the three requirements laid down in the will and the statutes: (1) that the works rewarded should constitute "discoveries," "inventions," or "improvements"; (2) that these should represent "the most modern results" and that "works or inventions of older standing ... be taken into consideration only in case their importance has not previously been demonstrated"; (3) that these should be works that "have conferred the greatest benefit on mankind." The use made of these criteria will be examined in the second section below.

DEFINING COMMITTEE PREROGATIVES IN THE MATTER OF PRIZE SELECTIONS

The commanding role of the committees in the matter of prize selections was apparent from the start, nowhere more so than in the form they gave their recommendations for the first year's prizes. Each committee presented the Academy with a single recommendation – a divided prize for Röntgen and Lenard in physics and an undivided one for van't Hoff in chemistry – backed up by short reports on the merits of these candidates. This set a pattern, and in the future the Academy would be provided with alternative recommendations primarily when members of the committees had been unable to agree.

Yet the stipulations in the special regulations (para. 7) that the committees "shall present to the Academy their opinion and proposals regarding the distribution of prizes" could be interpreted as indicating a less commanding role. Furthermore, those who favored an active role for the Nobel Institute had gained the right for the institute at the Academy to work alongside the committees "carrying out ... scientific investigation as to the value of those discoveries" proposed for the prizes in physics and chemistry (para. 12). In the first four years of the prizes, both the questions of the form that the committees' "opinions and proposals" should take and the role of the Nobel Institute were subjects of debate. How these matters were settled was an important factor in determining the future working methods of the committees.

If the Academy was to do more than simply ratify the choices of the committees, it had to be provided with information that would enable it to form an independent judgment about alternatives to the single prize recommendation of each committee. The initial demand of the Academy, put forward at the time of the first prize decisions, was a modest one: It only wanted to know the names of all the candidates put up for the prizes as well as those of the persons who had proposed them.[1] Since the chemistry committee had included the names in its prize recommendation of 1901, the Academy's request was aimed at the physicists who started to provide lists of nominees and nominators in 1903.

In 1902, the Academy became more ambitious and requested the committees to supply it with an overview of all candidates, comparing the merits of their work and giving the reasons why the ones proposed for the prizes should be preferred above all others.[2] This was more of a threat to the independence of the committees, since with more detailed information in hand about the alternatives to the single recommendation, that recommendation could more easily be challenged by the Academy. Predictably, the committees reacted negatively to the proposal. They asked the Academy not to take a formal decision that "would confine the freedom of action of its Nobel committees within narrower limits than those spelled out in the statutes."[3] If the Academy should feel obliged to act on its proposal, however, they requested that it be interpreted in the following manner. The detailed scientific analysis would as before be limited to the works recommended for the prize. It could possibly be extended, however, to another work if the merits of the latter were such that one would have to choose between the two. The committees were also prepared to give reasons for their elimination of candidates for the prizes of a given year, but would do so without ranking the works of these candidates. Instead, comparisons would be limited to the prospective prizewinner on the one hand and all the remaining candidates on the other. The committee proposal was adopted by the Academy and was put into effect in 1903.[4]

The Academy's decision provided the model for the committee reports that gradually became integrated into committee decision making. To comply with the Academy request, the committees supplied *general reports* where the works of all candidates were reviewed and the choice narrowed down to a select group from which were chosen the ones recommended for the prizes. Drawn up *after* the decision of the year's prize recommendation had been made, the general reports represented the summation of the decisions of a given year and a prognosis of those to be taken in the future. Although a grading of the candidates was avoided, the reports of the physics committee in particular generally contained a grouping of candidates based on the priorities that the committees attached to their rewarding. The reports also provided the opportunity to dismiss candidatures. Since this was most frequently done using the formal criteria of the will, we shall return to this later.

The Academy request for a detailed analysis of the works of the chief candidates led to the system of *special reports*. Commissioned from committee members, often the ones who supported the candidacy in question, the special reports initially only concerned the candidates that the committees recommended for the prize. With the growth of the number of leading contenders, however, and the lessening convergence of opinion among the nominators, the number of special reports tended to multiply. The important differences between the physics and the chemistry committees in the number and use of these reports was at least partly the result of the difficulty each committee experienced in instituting consensual decision making, and will be explored in that context.

Parallel to the discussion of how the committees should report to the Academy, the setting up of the Nobel Institute was being debated. Here, the practical problems of location, buildings, and budget predominated over the scientific ones of when and how the institute would investigate works proposed for the prizes; as it turned out, the final outcome was also determined by local and practical considerations.

In 1904, after the discussions about the Nobel Institute had dragged on for three years, Arrhenius announced that he had been invited by F. Althoff, a powerful figure in the Prussian Ministry of Culture, to stand for election to the Berlin Academy of Sciences. As described by Arrhenius, his position would be similar to that of van't Hoff, for whom a chair had been created at the University of Berlin at the same time as he was elected to the Berlin Academy.

Acting rapidly to avert the possibility that Arrhenius would be lost to Swedish science, the Royal Academy of Sciences decided to use part of the funds allocated for the future Nobel Institute to create what became known as the Nobel Institute for Physical Chemistry with Arrhenius as its director.[5] In this way, the wish that Arrhenius had expressed at the time of the negotiations over Nobel's will was realized; he was able to leave the Högskola and devote his time entirely to research as director of his own endowed institute, a position he was to occupy until his death in 1927. Obtaining this had not been easy, however, and his "call" to Berlin was not a guaranteed proposition, as Hasselberg was quick to point out. This led to a public confrontation between the two men from which Arrhenius emerged victorious; not only was the institute set up for him, but Hasselberg had to withdraw his accusations and give up the chairmanship of the Nobel Committee for Physics.[6]

The creation of an institute specifically for Arrhenius resolved the contradictions surrounding the role of the Nobel Institute in favor of the rationale that had loomed large in the minds of its proponents but that for obvious reasons could not be stated in the statutes. As shown in Chapter 3, Lindhagen and Pettersson had been primarily interested in a Nobel Institute for scientific research with strong links to the Högskola. True, Arrhenius's institute was a far cry from the "big" institute they had envisaged, but given the poverty of facilities for research in Stockholm, even the Nobel Institute for Physical Chemistry with its two staff positions was a welcome addition.

That the Nobel Institute's role in investigating works proposed for the prizes did not figure importantly in the discussions was probably due to the fact that in the years before the institute was established, the committees had discovered that they could dispense with an experimental verification of such works. Before the institute came into existence, the only recorded instance of experimental investigations was in 1903 when Ångström purchased one decigram of radium chloride for the physics committee, ostensibly to check the results of the Curies concerning the heat radiation of radium.[7] The results of Ångström's investigations were published in the journal of the Academy, and the sample also served to demonstrate the properties of radium in the lecture Ångström gave to the Academy prior to the meeting at which the decision was made to award half of the Nobel prize in physics to the Curies.[8] After the Nobel Institute had been established, Arrhenius was called on only three times during the period studied here to examine works proposed for the prizes, all of them involving the candidacy of Nernst.

The reasons for the hesitation of the committees to carry out their own scientific investigations are easy to understand. The experimental confirmation of most discoveries proposed for the prizes was of long standing. Theoretical works presented the committees with different problems, but even when experimentally based it would clearly have been beyond the capabilities of the Nobel Institute to carry out the experiments necessary to verify, for example, the Rutherford–Soddy disintegration theory or Nernst's heat theorem. In general, then, the information supporting the committees' recommendations for rewarding specific discoveries was obtained by culling the literature. As for the general overviews of the candidates' scientific careers and achievements, which occupied an important part of the special reports, these strongly resembled the reports submitted on candidates for professorial appointments or for membership in scientific societies.[9]

For the committees to carry out evaluations of candidates almost single-handedly required that the statutes be interpreted so as to facilitate their task. The first of these interpretations, chronologically speaking, was a prerequisite for the committees working out the several matters discussed above: It concerned the rights of the committees to meet jointly, a practice that was instituted in 1900 and approved by the Academy in 1901.[10] Initially these discussions were restricted to organizational matters; subsequently, they came to concern the question of how to evaluate candidates who had been proposed for the prizes in both physics and chemistry.

Probably as a result of the conflict that had been brewing since 1903, when both committees claimed the right to reward work in radioactivity, Pettersson proposed in 1908 (the year Crookes and Rutherford had been put forward for the prizes in both physics and chemistry) that the committees meet jointly to decide on the referral of candidates to one or the other committee. Although this proposal was accepted by the committees, the minutes do not show any formal decision to refer Crookes or Rutherford to either committee but only indicate that a discussion had taken place.[11]

Starting in 1910, however, the committees met jointly each year to decide on the referral of candidates who had been nominated in both disciplines.

To a certain extent, the decision to refer a candidate, and by implication the research area, to the physics or the chemistry committee depended on the formal boundaries drawn around the two disciplines. That atoms and elements were seen as the primary concern of chemists made it possible for the chemistry committee to claim works in radioactivity and atomic physics for *their* prize despite the fact that these subjects had far-reaching implications for physics. The specific candidacies involved represented a more important consideration, though, for unless there was strong support for these in one or the other committee, the research area might well be overlooked in the evaluations. That the referrals often depended on what each committee found expedient with respect to specific candidacies was most apparent, of course, in the case of Nernst. The frequent juggling of his candidacy between the two committees represents the primary example of how a candidate was claimed by each committee in turn for the purpose of negative rather than positive evaluation.

Another area in which the committees developed their own procedures largely without interference from the Academy concerned the nominations. The importance of the statutory right of committee members to nominate candidates for the prizes has been discussed in Chapter 4. If nominations by committee members were to influence deliberations, however, their authors would obviously have to be allowed to speak for these. That the committees felt obliged to make a formal ruling granting this right shows some influence of the view that they were to constitute juries whose members would be bound by rules concerning conflict of interest.[12]

More important, the rules concerning the nominations had to be "bent" so as to support rather than restrict the committees in their decision making. The main concern of the committees in this area was that the statutory rules not unnecessarily restrict the relatively limited number of nominations with which they had to work. Hence, the rule in the statutes that nominations be accompanied by the works referred to in these was interpreted liberally to mean that this was desirable but not necessary. Two other stipulations were relaxed for the same reason. The first concerned the requirement that scientists be nominated for specific works rather than more general achievements. The chemistry committee was more punctilious about this rule than the physics committee: In the former body, nominations were often disallowed on these grounds, whereas this was hardly ever the case in the latter. The second involved the principle that for nominations to be considered valid, they would have to arrive before February 1 of a given year and concern the prizes of that year. Here, the committees adopted the practice of considering the proposals that had arrived late as valid if they had been submitted by persons with permanent nominating rights. This practice began in 1903 when, as shown in the previous chapter, the physics committee hit upon this solution in order to include Marie Curie in the recommendation for the prize of that year. That the committees felt restricted in their work by the deadline of February 1 is shown by their petitioning the Academy in 1907 to change the rule so as to give them the right to decide, within one month after the end of the nominating period, which older proposals should be resurrected and added to the current ones. Although the Academy had never questioned the practice instituted by the Curie decision, it was not prepared to change the nominating rules formally, and the committees' request was left without action.[13]

THE USE OF STATUTORY CRITERIA IN COMMITTEE DECISION MAKING

In the early years, the committees not only established their prerogatives with respect to procedures but also with respect to the statutory criteria for choosing prizewinners.

In 1907, for example, the secretary of the Academy, probably with support from other members, requested that the committees justify their prize recommendations not just on the basis of scientific merit, but also with reference to statutory criteria.[14] The committees' answer to the Academy represented something of a rebuke. It stated that since they had always been concerned with the statutory criteria, nobody had ever been recommended for the prize whose work did not, in the committees' opinion, meet the requirements of the will.[15] The catch, of course, lay in the words "in the committees' opinion," for as long as the committees did not make explicit how the works recommended for the prize met the requirements that they constitute important discoveries, inventions, or improvements, that they be of recent date, and that they be those that had conferred greatest benefit on mankind, the Academy could not know how these criteria had been interpreted nor contest their uses.

As will become clear from the following discussion of each of these criteria, in general they were not used to pinpoint or even to justify the committees' choices, but constituted instead the overall standard to which some candidates measured up better than others, thus making it possible for the committees to narrow down the field. This standard was of course not immutable, but depended on the conditions governing the choice of any given year.

Discoveries, inventions and improvements

Considered individually, the specific meaning given each of the terms *discovery*, *invention*, and *improvement* determined its importance as a criterion guiding the committees to their choices. Here, the definitions of *discovery* were the most important; by contrast, those of *invention* and *improvement* played a subsidiary role.

The definition of *discovery* that most actively guided the committees was one closely akin to the popular sense of the word – that of finding something new, unexpected, and exciting. Placed in the context of the experimentalist orientations that predominated in the physics committee, discovery in the above-mentioned sense most often came to mean the detection of new physical phenomena and effects, particularly those associated with matter in its different forms, through experimental methods. That this definition came to play an important role in committee decision making (as shown, for example, by the awards made to Röntgen, Zeeman, Becquerel and the Curies, Rayleigh, and J. J. Thomson) was due not only to the experimentalist predilections of committee members but also to the fact that, at the turn of the century, the experimental approach was celebrating many triumphs: uncovering the new and unsuspected phenomena of X-rays and radioactivity; establishing that cathode rays consisted of electrified particles with a specific charge a thousand times that of the hydrogen ion, previously possessor of the largest known value of e/m; revealing that the atmosphere contained a hitherto unknown constituent, the gas argon; and so on. The committee was also fortunate in that these discoveries were all ready to be rewarded in the first years of the prizes.

In chemistry, the definition of *discovery* that most closely paralleled the one employed in physics concerned the isolation of new elements. Its impact on committee choices is borne out by the fact that almost all the scientists who had discovered major new elements between 1885 and 1905 were rewarded with Nobel prizes before the First World War: Moissan (fluorine), Ramsay (the noble gases), M. Curie (polonium and radium), and Rutherford (radioactive isotopes). In physics and particularly in chemistry, however, only a portion of the works proposed for the prizes fitted these notions of discovery. In these cases, the committees did not seek to develop alternative definitions, but instead made their choices most often through consensual decision making.

Even when the committees could pinpoint "discoveries," however, more specific notions had to be developed in order to implement the statutory requirement that the work selected be "the most *important* discovery." Committee reports shed some light on this matter.

For a discovery to be termed "important" it had to have been "processed"[16] not only in the sense of having been verified by other scientists, but also in the sense that its role in opening up new research areas should preferably have been realized. This is borne out in the following quotations from committee reports:

Investigations which have expanded our knowledge of the passage of electricity through gases with respect to a number of important problems and have opened up a new area of utmost value and importance for scientific research. (J. J. Thomson)

The high scientific value manifested as much by the scope as by the precision of these investigations without which the discovery of helium and the other so-called noble gases would surely not have occurred for quite some time. (Rayleigh)

The discovery of radioactive elements has opened up a new field of utmost importance and interest for physical research. . . . our fundamental view of certain questions in physical research has been altered and impulses have been given of a scope which we are not yet in a position to assess fully. (Becquerel and the Curies)

When used negatively, these same aspects formed the basis for the committees' judging a number of discoveries not "sufficiently important" to warrant a prize either because their "processing" by the scientific community had raised doubts about their validity (e.g., G. Le Bon's black light or R. Blondlot's N-rays[17]), or because, while perfectly valid, they did not seem to have the generalizability of "important discoveries" (e.g., J. Elster and H. Geitel's discovery of ionization as the source of atmospheric electricity, E. Goldstein's discovery of canal rays, and O. Lehmann's discovery of liquid crystals[18]).

If used actively in committee decision making, the terms *invention* and *improvement* would naturally have made for choices in the area of technology. That "pure" technology was represented in the period 1901–1915 by only two prizes – Marconi (1909) and Dalén (1912) – was due not so much to a lack of interest in such work by nominators and committee members (cf. Part II of this chapter) as to the difficulties that arose in applying these terms. The most important restriction put on the term *invention* concerned patented work, the rewarding of which had caused hesitation when the statutes were drawn up. Since no provision barring such work from the prizes had been included in the statutes, the physics committee took up the matter in 1901. It decided that a formal change was unnecessary, given members' agreement that, as a rule, "patented inventions should not be taken into consideration."[19] Proposals to reward patented work were not rejected out of hand, but in general the positive application of the term *invention* required – in addition to the considerations of its "benefit for mankind," to be discussed below – that the candidate be deserving of the prize because he had not derived any financial benefits from his invention. Few candidates met this requirement, since if an invention had been successful, it had normally been patented and hence brought some wealth to its author. That Marconi was not yet the wealthy man that he was later to become, and that Dalén could be considered truly deserving having lost his eyesight in an accident involving experimental work shortly before the award, probably facilitated their receiving their prizes.

The many difficulties involved in translating the statutory terms into specific criteria for choice should not obscure what was perhaps the most important function of these terms in committee decision making: Taken together, the terms *discoveries*, *inventions*, and *improvements* provided the main rationale for the committees to restrict their evaluations and recommendations to specific works rather than to the more general achievements or the life work that was frequently cited in the nominations. Proposals of the latter kind could thus be dismissed. As already noted, in chemistry they were sometimes not even reviewed, whereas in physics proposals were discarded out of hand only when there was no mention whatsoever of the grounds on which the person should receive the prize. But the lack of reference to specific works was not in itself a ground for dismissing the *candidacy* (meaning thereby both the person and his work), since if the committee was willing, such works could always be identified at the evaluation stage. The requirement that specific works be cited was used irrevocably for persons whose life work was clearly drawing to an end; thus the candidacies of such older statesmen of science as M. Berthelot and Lord Kelvin were turned down despite considerable nominating support. In other cases, committee standards were much less firm and depended on the overall standing of the candidacy. For instance, in 1905, when some members of the chemistry committee felt that the time to reward von Baeyer had come, they cited among other arguments that von Baeyer had previously always been cited for his life work but that this had now changed since he was being proposed for specific discoveries.[20] The same "changed circumstances" were invoked in favor of Ostwald in 1909.

Recency of achievement

The rather vague meaning given the requirement of recency in the statutes made it difficult for the committees to set standards in this area and, above all, to impose these on the nominators. Other prestigious awards, such as the medals of the Royal Society of London, were commonly bestowed on older scientists whose achievements belonged to the past. The replies to the invitation to nominate candidates for the Nobel prizes in the first year indicate that the nominators were puzzled by the requirement that the discovery to be honored should be of recent date; Warburg, for instance, expressed the opinion that while this may be preferable, he did not see that it should be necessary.[21] The committees' dilemma was compounded by their being faced with a backlog of scientists whose work had received widespread recognition in the latter part of the nineteenth century and whose rewarding would definitely lend glory to the newly established prizes. Arrhenius summed up the situation when he wrote to a friend in 1908: "We shall go on for some more years having good prizewinners in physics and chemistry, that is, those who are recognized by everybody.... after that I hope things will develop as Nobel had probably envisaged, namely that young men who have done good work shall receive the prize even if some old fogeys are against it.... The worst thing would be," he concluded, "for the prizes to develop into an old-age pension."[22]

In resolving these problems, both committees preferred to err on the side of conservatism. Under the general standards applied for recency, works conducted within the past two decades were considered for the prizes. With a steady supply of works dating back no more than ten years from the rapidly moving field of "ray" physics (i.e., X-rays, cathode rays, radioactivity, and magnetooptics), the physics committee could permit itself to be choosier than the chemistry group, as shown by its more frequent dismissal of investigations as being "too old." Younger scientists also won more prizes in physics than in chemistry; eight out of the first twenty physics laureates were under forty-five years old, as compared with five out of sixteen in chemistry. For candidates not to be too old may also have been a factor that helped overcome the fact that the work for which they were honored dated back more than a decade (e.g., van't Hoff and Arrhenius) or was carried out over a long period of time (e.g., T. W. Richards). In the case of the work of older candidates – von Baeyer and van der Waals, for instance, who were both in their seventies – it was important that they had continued these into recent times, even if no significant breakthroughs had occurred. Others in this category (e.g., Cannizzaro, Curtius, and Righi) were turned down, however, on the grounds that their achievements belonged to a previous era. Here, committee judgments would most often depend on the competition; when this was as strong as in the field of radioactivity, for instance, the older men would have to stand back, as shown by the chemistry committee's favoring Rutherford over W. Crookes.

The case of Crookes, who was over seventy when he was first proposed for the chemistry prize, illustrates well the committee's dilemma in implementing the requirement of recency. On the one hand, there was Ramsay's argument (taken up by Pettersson) that the rewarding of Crookes (who had done pioneering work not only in the field of radioactivity but also in "ray" physics in general through his method of creating high vacua) should take precedence over that of his followers.[23] On the other hand, there were the considerations of recency and importance of work that made the candidacy of Rutherford prevail over that of Crookes. Although not invoked in the case of Crookes, a rationale for rewarding pioneering research was present in the part of para. 2, which states that older works could be considered if their importance had not previously been demonstrated. This was the basis on which the chemistry committee recommended D. I. Mendeleev in 1906 for his periodic law of the elements, which dated back to the 1860s and early 1870s. The rationale given was that the recent discovery of the inert gases argon and helium by Rayleigh and Ramsay had shed new light on Mendeleev's system by suggesting a group of zero valency. When Mendeleev's candidacy (which was strongly supported by Pettersson) was discussed in the chemistry section, K. A. H. Mörner, who had participated in drawing up the statutes, contested such a use of para. 2, since in his opinion Mendeleev had not predicted or even foreseen a 0-group in the table. The committee's recommendation was defeated, but only by a slight margin: Mendeleev received four votes against five for Moissan.[24]

The Mendeleev case illustrates well the difficulties involved in using para. 2. These occurred chiefly because "pure" rediscovery of the Mendelian kind is rare in the history of science. Instead, the revaluation of older works most often comes about through the reawakening of interest in a given research area, which in turn produces more recent and – in the Nobel context – also more prizeworthy discoveries. Here again, the final outcome would depend on the competition and on committee priorities. When the physics committee needed an alternative to Ångström in 1910 – in case a posthumous prize would not be acceptable to the Academy – it recommended van der Waals, whose work on the corresponding states of gases and liquids had not previously been considered of sufficient recency, dating back as it did to the 1870s. With growing committee and nominator interest in the rewarding of Kamerlingh Onnes for his liquefaction of helium, achieved in 1908, the work of van der Waals, which had provided the theoretic basis for Kamerlingh Onnes's investigations, took on new importance, and his award served as the wedge for that of Kamerlingh Onnes.[25] The same factors – competition and committee priorities – influenced, in this case negatively, the rewarding of Boltzmann, whose work in kinetic theory the physics committee pronounced as being "too old."[26] For the committee to have realized the value of this pioneering work in statistical mechanics would have required a readiness to award a prize to Planck. This was not yet the case, however, when Boltzmann died in 1906.

For the benefit of mankind

In the absence of guidance from the statutes on how to interpret Nobel's prescription that the works rewarded be those that "have conferred the greatest benefit on mankind," there was no basis for the committees to translate this requirement into specific criteria for choice. For the most part, this provision did not operate as a standard against which the candidates were measured but, rather more remotely, as a reminder to the committees not to stray too far from the course Nobel had laid down for the awards. It was only when work in technology was being considered that the utility of inventions or improvements became a criterion for choice. In most other cases, "benefit of mankind" was taken to mean that it was preferable if the work rewarded had some utilitarian value.

Being academic scientists, most committee members probably subscribed to the notion that if the prizes served to foster scientific progress, they would benefit mankind. They probably also felt that it was well nigh impossible to predict or to pinpoint the multifarious and diffuse applications that might flow from a discovery in basic research. If the committees were not particularly prone to consider utilitarian aspects in their prize evaluations, the situation in the Academy was different. As early as the first award year, the Academy felt compelled to assert its authority in this area. As Phragmén told Mittag-Leffler when reporting on the plenary meeting held to decide on the prizes of 1901, the exclusion of Lenard from the final award decision in physics, despite his having been recommended by both the committee and the section, was largely the result of the argument that Lenard's work, although a precursor to Röntgen's, had not conferred any "benefit on mankind." Furthermore, to counteract similar objections with respect to the chemistry prize, Arrhenius had been obliged at the last minute to prepare a special memorandum on van't Hoff's work that highlighted the *practical* applications of solution theory, particularly in the medicophysiological realm.[27] It seemed to Phragmén that "benefit of mankind" would be both the "catchword and the divining rod" for the future.[28]

Here he was wrong, but the experience of the first year still made the committees more careful about including references to the utilitarian value of the works they recommended for the awards. Finding such a rationale was not excessively difficult. In chemistry, the interpenetration of academic research and industrial research and development meant that a large number of the discoveries coming out of university laboratories eventually found some use in industry. Some examples: Fischer's work on sugar and purines and that of von Baeyer on indigo were of obvious use in the food and dyestuffs industries, respectively; Sabatier's method of hydrogenation of organic compounds found important applications when developed into the process of "fat hardening" used in the manufacture of stearin, soaps, and vegetable oils; the electric furnace developed by Moissan led not only to the isolation of fluorine (cited in his award) but also to the laboratory preparation of many metals and carbides, among them artificial diamonds, or so it was thought at the time.[29] In physics, utilitarian or, better yet, humanitarian considerations were of course most prominent in the rewarding of Röntgen. Such a rationale could also be found for awards in the field of radioactivity, since experiments with radium as a therapeutic agent against cancer had begun at the time the Curies received their prize. In most other cases, excepting technology, the committee was hard put to point out practical applications of the works in basic physics that they were most often recommending for the prizes. Perhaps sensing opposition in the Academy, committee members nevertheless made reference to such applications, as for instance when Arrhenius in his special report of 1908 pointed out the practical utility of Planck's radiation theory for illumination engineering and techniques.[30]

The most direct influence of Nobel's wishes came about through the inclusion of technology among other research specialties, albeit not as a very important one, as an area for prospective prize awards. When a choice was made between different technological innovations on the basis of their practical value, the term *benefit of mankind* was most often used in a way that reflected the utilitarianism of Bentham and Mills; that is, for an innovation to be seriously considered for the prize, it should have provided a substantial amount of good to a great number of people. If commodities were involved, the question of whether or not these could be produced on a large scale and for a reasonable price became a standard. The same criterion could clearly not be applied in the area of "collective goods," but the physics committee rewarded the development of wireless telegraphy, which had already proven of great worth to a large number of people, yet refused to reward the invention of the airplane because of the loss of life it might entail; this shows the influence of the same philosophy.

The increased attention given the field of technology by both committees in the years around 1910 was largely due to dissatisfaction on the part of the technologists in the Academy, who felt that the awards had come to reflect an overly academic research orientation. In 1912, the plenary meeting asserted its authority as the final arbiter in a more striking manner than in 1901. The recommendation of the physics committee and the physics section for a prize to Kamerlingh Onnes was voted down by a substantial majority. Instead, Dalén, the Swedish inventor, received the award "for his invention of automatic regulators for use in conjunction with gas accumulators for illuminating lighthouses and buoys." Having won what they considered an important victory, the techologists in the Academy do not seem to have made any further attempts to influence the prize decisions, and the concerns of the committees with awards in the area of technology also diminished.[31] To the physicists in particular, these concerns had always been subsidiary to the preeminent role assigned experimental physics, which, as will be discussed in Part II of this chapter also determined the committee's views of what were "the most important discoveries" in physics.

Since neither the formal statutory criteria, nor the Academy's occasional assertion of its rights with respect to these, nor even the nominating system provided the committees with unambiguous guidelines for selection, how then did the committees arrive at their choice of prizewinners? This question will be addressed next.

Part II: The role of consensus

By looking at some inherent structural features of Nobel prize decisions, we can discern three key elements in committee decision making: (1) Since many more scientists were proposed for the prizes than could possibly win them, the final decision was the result of a *choice* among such a large number of alternatives that rational decision-making procedures (e.g., weighing the merits of each case and the consequences of each action) did not apply; (2) the institution was best served if, in retrospect, the award decision appeared as the only possible course of action, and so the selection procedure had to be aimed at providing *justification* for the final choice; (3) the narrowing down of alternatives to one final and seemingly inevitable choice was most effectively done if there was *consensus*.[32]

Among these elements, consensus proved to be decisive when it came to the final choice. Consensus can be defined as the degree of personal commitment that participants feel toward a group decision after it has been reached.[33] As more than just agreement, it represents a form of ex post facto validation of a "good" choice.[34] This was particularly important in committee decisions where the choice had to be justified and spoken for at the Academy. Consensus was important not only at the stage of justification, however, but was sought throughout the decision-making process. In fact, it was an integral part of the selection of prizewinners because it directed the committees' attention to candidates who would be acceptable to the majority of the members. We shall now examine some general features of consensual decision making in two committees. Having established that the committees differed significantly in their ability to achieve consensus, the specific reasons for this will be discussed.

AN OVERVIEW

"L'unité fait la force" was how Söderbaum characterized the strategy that the chemistry committee should follow in 1905.[35] There is no doubt that this maxim sums up an essential feature of both the proceedings leading up to the committees' choice of prizewinners and the acceptance or rejection of this choice by the Academy. The two are intimately related, for a recommendation adopted unanimously and supported by the committee members in the Academy would usually be ratified, whereas if the members could not agree, the decision would be thrown into the sections of physics and chemistry, or the Academy meeting in plenary session, with generally unpredictable results.[36] Consensus was thus the main mechanism for safeguarding both the independence of the committees vis-à-vis the Academy and the orderly functioning of the process of prize selection.

But apart from the functional importance of consensus, which when extended outward shielded the institution from the loss of prestige that would have ensued if there had been public conflict, it was a feature valued in itself. It is not possible here to elaborate on the role of consensus in Swedish society; it suffices to say that the country's bureaucratic system, built as it was on collegia (and nowhere more so than in the universities) possessed elaborate rules for bringing about unanimous administrative decisions.[37] Several of these (e.g., voting procedures and the use of dissenting opinions) were carried over to decision making on the Nobel prizes. Whether prompted by valuational or utilitarian considerations – the two are difficult to separate – the committees' striving for consensus was important, chiefly because it represented the main form of rationality in the choices (i.e., through consensual validation) but also because, when there was hesitation about the choice, the candidate most likely to be selected was the one on whom the majority of the members could agree.

Studies of small-group decision making have shown that the process whereby preliminary ideas are transformed into consensual decisions can be divided into several distinct phases. Labeled according to the type of interaction that predominates during each phase, these have been described as orientation, conflict, emergence, and reinforcement.[38] Applied to committee decision making, our earlier analysis of the actions of Arrhenius and Mittag-Leffler concerned primarily the first two phases. By comparing the decision making in the physics committee with decision making in the chemistry committee, it is possible to shed some light on the two latter phases of emergence and reinforcement.

The matter of *when* the final decision on the recommendation of the year's prizewinner emerged in each committee is revealing because, not surprisingly, this depended on the members' ability to agree. When the process was flowing easily the choice could be made as early as the month of May, but when complications arose the final decision would be delayed until the end of September, the statutory deadline for the committees' submission of recommendations to the Academy. The physics committee was almost always able to agree on the final choice in the spring or, at the latest, when the committee met for the first time after the summer recess. The exceptions all represent occasions when the committee was uncertain about the final choice: Planck (1908), van der Waals (presented as an alternative to Ångström in 1910), and Wien (substituted for Gullstrand in 1911). The chemistry committee, by contrast, hardly ever reached a decision in the spring. Here, the final choice often did not emerge until the very end of the statutory period. It is also noteworthy that the chemists experienced difficulties in arriving at a prize recommendation not only in the years when nominators' opinions did not converge around any particular candidate – that is, after 1907 – but also before then.

Differences in the ability of the two committees to arrive at a decision are also reflected in the amount of information formalized in special reports that was brought into the selection process. In the great majority of cases, the physics committee reached its decision on the basis of the nominations and the members' informal general knowledge of the candidates. It was only *after* a decision had emerged, that is, in the reinforcement stage, that one of the members was commissioned to draw up the special report that contained the committee's motives for its prize recommendation.[39] Reports on more than one candidate were usually drawn up only when there was disagreement over the final choice; for example, in 1908, the committee commissioned reports on both Lippmann and Planck. These practices in themselves do not provide strong evidence of a greater ease in arriving at a final choice, but their significance in this respect becomes clear when compared to those of the chemistry committee. Here the practice, instituted after the first four years, of commissioning reports on several leading contenders was eventually extended to all candidates whose names could be expected to reappear in the nominations. When a new candidacy was put before the committee, a report was commissioned early in the decision-making process. Reports on "older" candidacies would sometimes be added to those on the new ones, with the former being resubmitted in their original form or supplemented with new information.

That the choice that emerged from the chemistry committee was based on a greater amount of knowledge formalized in special reports may indicate that the chemists were more meticulous about the selection of prizewinners than the physicists. In important

respects, however, the proliferation of reports in chemistry reflected the divergence of opinion that reigned in the committee,
since members would frequently argue the case for the candidates
they favored in the special reports. Reports on given candidates
often constituted *decision proposals* by individual members, for in
numerous cases these members were also the ones who had *nominated* the candidate in question.

Finally, the greater difficulties in formulating a recommendation
experienced by the chemists are reflected in those instances when
the committees had to vote on the final choice. In chemistry, this
occurred in six out of the first fifteen years.[40] In physics, by contrast, the choice only once went to a vote – in 1911, between
Gullstrand and Poincaré. In this case the choice was a clear-cut
one, with Carlheim-Gyllensköld alone favoring Poincaré. In the
chemistry committee, however, the voting often started out with
three or even four proposals that had to be narrowed down in
successive rounds.

Why did the chemists experience more difficulties in their decision making than the physicists? Looking at some of the general
conditions under which the committee had to make its choices,
one finds that they differed from the physics group in three important respects.

First, the practice of not dividing the chemistry prize quite naturally restricted the committee in its margin of maneuver. It is
not known how this custom developed, only that it was observed
throughout the period studied here. As discussed in Chapter 4, it
was also reflected in the nominations. In the one exception to this
rule – the prize for Grignard and Sabatier in 1912 – special arguments for the necessity of such an unusual step were developed
by Widman in the special report he wrote on the joint candidacy.[41]
The difficulties that the practice of undivided prizes caused the
committee in its decision making were particularly keenly felt with
respect to inventions or improvements in chemical technology,
which were often the result of collaborative or competitive efforts
(see the section below, "Divergence in chemistry").

Second, the physics committee was more fortunate than the
chemistry committee in benefitting from strong leadership. Although the statutory power of the chairman was limited to casting
the decisive vote in case of a tie (i.e., when only four members
were present), his real power could be much greater – depending

on the respect he commanded from his fellow members, on his ability to plan ahead for a choice that would be acceptable to the committee, and on guiding the committee toward that choice. There is no doubt that Ångström and Granqvist, who chaired the physics committee during the major part of the period studied here, were instrumental in bringing about the consensus that generally surrounded decisions in the committee.[42] The assessment of Ångström's chairmanship by a fellow member (H. H. Hildebrandsson) reads like a study in the social psychology of group leadership, for he lists many of the traits commonly associated with "good" chairmanship: "unsurpassed knowledge of all areas of physics," "strength and clarity of. . . argumentation," and most important, "he did not *à tout prix* press his own opinion; . . . when there was a divergence of views he sometimes adopted someone else's opinion. . . in order that consensus be achieved."[43]

No such consistently strong leadership was exerted over the proceedings in the chemistry committee. The first chairman, Cleve, occupied this position only three years (until his death in 1904). Although his knowledge of different fields of chemistry was undoubtedly the greatest of any of the members, his leadership may have been impaired by his minority view on the question of the prize for Arrhenius. Pettersson, who succeeded him, could not spend much time on committee matters because of his heavy involvement with the hydrographic research station he had founded in western Sweden. He resigned from the chairmanship in 1910, and in 1912 his colleagues asked for his resignation from the committee.[44] Although Pettersson tried to anticipate the committee choice through decision proposals at the nominating stage (for example, in the cases of D. I. Mendeleev, F. A. Kjellin, and G. de Laval), it was only with difficulty that he won a committee majority for the first of these. His actions may therefore have had a divisive influence. It was not until Hammarsten took over in 1910 that the committee acquired a reasonably strong chairman. His task was not an easy one, however, for by this time achieving consensus had been rendered very difficult through yet another condition specific to decision making in the chemistry committee.

This, the third difference between the two committees, concerned the degree of loyalty that the members developed toward the group and their willingness to work together as a team, perhaps the most essential element of all. The physics committee was aided in this respect in that its majority was being made up of Uppsala physicists working in the experimentalist tradition. The membership of the chemistry committee represented a range of specialty orientations – organic chemistry (Widman), agricultural chemistry (Söderbaum), physiological chemistry (Hammarsten), and chemical technology (Klason), for example – and was therefore more fragmented.

It is now in order to look in more detail at how differences in the specialty orientations represented on the committees influenced their relative abilities to reach consensus about the choice of prizewinners.

THE SPECIFICS OF CONSENSUS

Convergence in physics

As noted in Chapter 4, from 1907 until the end of the period studied here, the physics committee's decisions were frequently at variance with majority "votes" among the nominators. That the committee could disregard such votes and still arrive at a consensus was due to shared beliefs in the primacy of the experimental approach to physics. More specifically, the deliberations in the physics committee show the two main mechanisms through which work in experimental physics was given precedence: because this was the specialty orientation of the committee majority, and because work in experimental physics closely conformed to the statutory requirement that the prizes honor "discoveries." The situation of rapid disciplinary change that affected physics in the first decade of the century meant, however, that such a restricted view became problematic. Hence, the committee majority's definition of physics, which favored the experimental side of the discipline, had to be defended against incursions from "other" fields, primarily mathematical and theoretical physics.

The broad base for consensual decision making in the committee was the view that the experimental approach was the key to scientific achievements and advances. This view was widely shared among physicists at the time, and was reflected in the nominator support given candidates whose work met the committee's definition of "important discoveries" in physics (cf. Part I). While the rewarding of such work presented no problems, the committee

Selected members of the Nobel Committee for Physics elected 1901–
1915: Top left: Knut Ångström (1857–1910), professor of physics at the
University of Uppsala. (Courtesy of the Uppsala University Library)
Top right: Vilhelm Carlheim-Gyllensköld (1859–1934), professor of
physics at the Stockholm Högskola. (Courtesy of the Royal Library)
Bottom left: Gustaf Granqvist (1866–1922), professor of physics at the
University of Uppsala. (Courtesy of the Uppsala University Library)
Bottom right: Bernhard Hasselberg (1848–1922), head of the Institute
of Physics at the Royal Academy of Sciences. (Courtesy of the Uppsala
University Library)

Selected members of the Nobel Committee for Chemistry elected 1901–1915: Top left: Per Theodor Cleve (1840–1905), professor of chemistry at the University of Uppsala. (Courtesy of the Uppsala University Library) Top right: Otto Pettersson (1848–1941), professor of chemistry at the Stockholm Högskola. (Courtesy of the Royal Library) Bottom left: Henrik Gustaf Söderbaum (1862–1933), professor of agricultural chemistry and head of the Experimental Research Station at the Academy of Agriculture. (Courtesy of Uppsala University Library) Bottom right: Oskar Widman (1852–1930), professor of chemistry at the University of Uppsala. (Courtesy of Uppsala University Library)

was soon to learn that important discoveries were few and far between. However, this did not absolve them of the obligation to recommend a prizewinner. That they were able to do so without relying on majority votes among the nominators was due in the first instance to the strong influence of the Uppsala school in the committee. As noted in Chapter 2, members of this school were "experimenticists" in the sense that they attached high value to the construction and use of instruments for precise measurements. This orientation was also reflected in the prize recommendations.

The most telling example of this interest in instrumentation is the prize awarded A. A. Michelson in 1907. In the history of science, Michelson is best known for conducting the Michelson–Morley experiment (1887) that disproved the hypothesis that the velocity of light is dependent on the velocity of the earth moving through a fixed medium (the ether) – an experiment that has been seen (incorrectly, as Holton has shown) as underlying Einstein's special theory of relativity.[45] However, when Hasselberg set out to "procure the prize for Michelson," as he stated in a letter to G. E. Hale (who had nominated Michelson), it was as "an opinion of sympathy for an area closely connected with my own specialty" and because, he exclaimed in another letter to Hale, "I cannot but prefer works of *high precision.*"[46] The specific elements of Michelson's achievements that evoked Hasselberg's admiration were the applications of his major instrumental innovation – the interferometer – to spectroscopy and metrology, research areas especially close to Hasselberg's heart because of his own investigations and because of his work as the Swedish representative on the International Committee of Weights and Measures (see Chapter 2). Michelson's work with the interferometer to determine experimentally the length of the international meter introduced the notion that metrology could be based on natural constants. Furthermore, Michelson's interferometer had been used to recalculate the wavelength tables for spectroscopy. It is significant that these were in Hasselberg's opinion the most important aspects of Michelson's work. The lengthy treatment of these achievements in Hasselberg's special report on Michelson stands in marked contrast to the one-sentence reference to the Michelson–Morley experiment.[47]

The admiration of Hasselberg and (although less well documented) of other members of the Uppsala school for Michelson's work in precision physics was sufficiently strong to overcome the fact that Michelson was proposed by only two people for the year in which he received the prize, and had been proposed only once before then (by E. C. Pickering, the Harvard astronomer). Moreover, Michelson's work did not contain any specific element that could be heralded as a discovery. Hasselberg's awareness of this possible conflict with the statutes is borne out in a letter in which he asks Hale to suggest which of Michelson's works might constitute a discovery.[48] In the end, Michelson's award citation mentioned no discovery but referred instead to "his optical precision instruments and the spectroscopic and metrological investigations carried out with their aid."

There were other occasions when works in precision physics were the subject of favorable evaluations, but only in one case did these result in a prize recommendation. In 1910, Knut Ångström was recommended for his compensating pyrheliometer, which had led to improved measurements of the solar constant. As noted in Chapter 5, the Academy preferred not to make a posthumous award.

Predilections for precision physics and instrumentation also made the committee majority open to candidates in fields other than what Hasselberg had termed "pure Physics properly so-called" (i.e., experimental physics).[49] Astrophysics was a promising field for an award, since here the methods and instruments of experimental physics were being successfully employed to extend the boundaries of physical research beyond the earth and the atmosphere. Solar research was the fastest-moving field in astrophysics, due mainly to instrumental innovations, among them the spectroheliograph, which brought a number of phenomena concerning the sun and its atmosphere within the reach of more precise observation and more adequate explanation.

The spectroheliograph, developed by G. E. Hale, had been used at the Mt. Wilson Solar Observatory in California to investigate the distribution of substances in the solar atmosphere. This work led, among other things, to the discovery of magnetic fields in sunspots. At about the same time as Hale, H. Deslandres had independently invented his version of the spectroheliograph at the Paris Observatory. Here too it was employed to elucidate the complex phenomena of solar activity.[50] Of the two men, Hale was the first to be proposed for the prize (in 1909); the proposals cited his instrumental innovations and the investigations carried out with these. He was immediately put in the select group of candidates whose work, in the opinion of the committee, particularly merited the prize, although, as Arrhenius wrote in the general report of that year, "the novelty of these investigations makes it advisable not to fix upon them now."[51] In 1913, when both Hale and Deslandres were proposed, although not for a joint award, Hasselberg noted in the general report that their endeavors ought to be followed very closely, since "without doubt, they would sooner or later very rightly come to be regarded as deserving the physics prize."[52] Again in 1914 and 1915, they were considered among the very top possibilities, but because only Hale was nominated in the former year, the committee decided to delay a detailed evaluation of their work. In 1915, success seemed close at hand, for after Arrhenius had nominated *both* Hale and Deslandres, Hasselberg drew up a special report in which he was predictably favorable to their receiving the award. An unfortunate circumstance intervened, however. Carlheim-Gyllensköld favored his own candidates in astrophysics, K. Birkeland and C. Störmer, whose work on the aurora borealis was the subject of his nomination–cum–special report. That three members of the committee favored an award in this field yet were unable to agree on who should receive it was probably the reason why the matter was left in abeyance. When it was taken up again after the war, the chances for an award in the area dwindled as the committee switched its priorities to theoretical and atomic physics.[53]

Given the primacy that most members put on experimental work, it was natural that they should consider theory construction to be intimately bound up with experiment. The desire to put theory and experiment on the same footing even when they were not joined in the work of one person is present in the coupling of Lorentz's name with Zeeman's in the awards of 1902 and in the opinion expressed by the majority in 1908 that if Planck were to be rewarded, the prize should be divided between "the foremost theoretician and the foremost experimentalist [i.e., O. Lummer]."[54]

When the committee was moved to reward more theoretical work – that is, that of the Braggs and von Laue – toward the very end of the period studied here, it chose theories whose development had involved close interaction with experimentation. Furthermore, this had generally come about by the physicists' assuming alternately the role of experimenter and theoretician, a combination far from uncommon at a time when theoretical physics had not yet achieved the status of an independent specialty. Consider a specific example: the work for which von Laue received the prize in 1914. In 1912, one month after von Laue and two experimental physicists (W. Friedrich and P. Knipping) reported jointly on the diffraction phenomena recorded when crystals were irradiated by X-rays, von Laue alone submitted a publication giving an explanation of the phenomenon in the form of a quantitative theory that subsumed X-rays under light waves.[55] When the physics committee awarded von Laue the prize two years later, it was on the basis of its "unanimous opinion" that this was "a most beautiful and significant work from the point of view of both theory and experiment."[56]

Not surprisingly, it was the awards to Wien and Planck for their successive radiation formulas that put the committee's conception of theory formulation to its most severe test. It is also significant that Arrhenius, who was more open to new developments of physical theory, should have broken ranks with the committee majority over the question of an award for Planck. Planck's radiation law of 1900 satisfied the requirement of experimental verification. Furthermore, it made possible the independent verification of known constants (the use of k to calculate Avogadro's number) as well as introducing the new constant h.[57] Yet, as the committee was to learn in 1908, the assumption of the quantization of energy that Planck had had to introduce (much against his will) in order to derive his law was so far-reaching as to amount to a general theory[58] that represented a break with the map of the physical world. If the awarding of Planck was left in abeyance, then, that of Wien (who received the prize in 1911) presented fewer difficulties, for as had already been pointed out in the Academy debates of 1908, his displacement law was a strictly logical consequence of established laws in physics. However, the nominations for Planck continued to accumulate. As the successive evaluations of Planck's work moved from radiation formulas to quantum theory, the committee, lacking expertise in the matter, also had to rely more heavily on outside authorities (e.g., H. A. Lorentz, Wien, H. Abraham, and J. Stark).[59]

The committee's absence of expertise in mathematical and theoretical physics had not been felt at first, since initially these fields were given low priority as areas for awards. This was shown by the attitudes of the Uppsala physicists when the election of a mathematical physicist to the committee was first debated (1902–1904). This was not an urgent matter, they argued, since "an investigation in mathematical physics would only rarely be considered for rewarding unless, as was the case...for Lorentz's electron theory, it leads to 'discoveries and inventions' in the area of experimental physics."[60] This categorical statement was issued in response to Arrhenius's equally forceful statement that the absence of a mathematical physicist on the committee in the year Lorentz was awarded the prize constituted a "very serious deficiency." It was necessary to fill this gap rapidly, he argued, since several proposals requiring competence in mathematical physics were already in hand; failure to do so would throw serious doubt on the committee's ability to evaluate work for the prizes.[61] Neither Arrhenius nor the Uppsala physicists were arguing in the abstract, of course, but each side was hotly opposing the other's candidate for the committee seat left vacant by Thalén. In launching an all-out campaign for Vilhelm Bjerknes, professor of mechanics and mathematical physics at the Högskola, Arrhenius was motivated as much by his wish to have someone of Bjerknes's stature on the committee as he was to secure him for Swedish science by somewhat improving his material conditions. Arrhenius's campaign notwithstanding, the Uppsala candidate, G. Granqvist, was elected by a small majority (sixteen votes against thirteen for Bjerknes).[62] It was unfortunate that the deteriorating relations between Sweden and Norway prevented Bjerknes, a Norwegian,[63] from being elected to the committee, since his knowledge of mathematical and theoretical physics would have brought the committee badly needed expertise in these fields.[64]

The need for representation of these aspects of physics was again the issue in the elections of 1910, when the seats left vacant by Ångström and Hildebrandsson had to be filled. By then it was no longer possible to rule out awards in this field, for nominations were definitely on the increase; in 1910, it should be recalled, Mittag-Leffler was flooding the committee with nominations for Poincaré, although the more portentous event of that year was the first nomination of Einstein (by Ostwald) for his special theory of relativity. Mathematical and theoretical physics were also on the rise in Swedish physics. Whereas in the earlier election the choice had been, to quote Ångström, between "the theoretical–

mathematical physicist or the more practical, theoretical electrician,"[65] there were now three candidates who all represented mathematical or theoretical physics in different ways.

It is significant that the two who were eventually elected – Carlheim-Gyllensköld and Gullstrand – were physicists who had gained competence in mathematical physics both through their work in relatively mathematized research areas (geodesy and geometrical optics, respectively) and through avocational interests. In both elections, the loser was Ivar Fredholm, Bjerknes's successor at the Högskola, whose superiority as a mathematical physicist nobody would deny but who was perceived by some members of the physics section as a mathematician – and one of Mittag-Leffler's "troops," to boot – lacking the necessary knowledge of physics.[66] In opposing Fredholm, the rank-and-file physicists were reacting to what they probably perceived as a threat from a rising specialty by drawing the boundaries more tightly around the domain they defined as "physics proper." By contrast, those at the forefront of disciplinary development were not afraid of expanding or blurring the definitions of physics so as to avoid the chasm that might otherwise open up between "physics proper" (i.e., experimental physics) and mathematical and theoretical physics.[67]

However, Carlheim-Gyllensköld soon found that such a gulf existed when he tried to convince the committee that more awards should be made in mathematical physics. The occasion was his dissenting opinion in favor of Poincaré in 1911, in which he discussed the "matter of principle" of awards in mathematical physics. His analysis of the first ten years of the prizes pointed up the de facto exclusion of works in this specialty. With the exception of H. A. Lorentz and, to some extent, J. J. Thomson, he stated, "the Nobel prizes have up to now been awarded to experimental physicists." He stressed that the neglect of mathematical physics was not due to an absence of nominations, for prominent representatives of this specialty (e.g., Boltzmann, Duhem, Heaviside, Kelvin, Poincaré, Wien, Planck, and Einstein) had often been put forward for the prize. In most instances, these nominations had been made by experimental physicists: Lenard, Becquerel, Röntgen, Zeeman, J. J. Thomson, to mention only the first handful of names in the long list he presents. "This shows clearly," he wrote, "the meaning that the scientific world attaches to the term 'physics' and what it considers as falling within the domain of a prize in physics. Justifiably, these votes merit attention."[68]

It was unfortunate that the majority of nominators' votes in mathematical physics should have concerned Poincaré, whose candidacy, for the several reasons discussed in Chapter 5, could not garner a majority of committee votes. Rather than opening up physics to mathematical physics, the campaign for Poincaré, given that both the candidate and his supporters were so closely identified with the world of mathematics, caused a retrenchment in the committee. Arrhenius, for example, reacted by reasserting the experimentalist philosophy; his claim that a theory would have to be confirmed by experiment in order to qualify for the prize was both unrealistic in the case of Poincaré's cosmical physics and inconsistent with his actions on behalf of other candidates.[69] As for the Uppsala physicists, they focused their efforts on procuring the prize for two of their own (Ångström and Gullstrand, consecutively.) A minor victory had been won, though, in that the majority of the committee conceded that mathematical physics was part of physics and hence qualified for consideration.

In committee decision making, the campaign for Poincaré marked the end of the period when the specialty of mathematical and theoretical physics was conceived both in terms of higher mathematics applied to phenomena in nature and with reference to physical theories expressed in advanced mathematical language. By 1912, this bond had been broken, and of Carlheim-Gyllensköld's long list of candidates in mathematical physics only three serious contenders, all theoretical physicists, remained: Planck, Nernst, and Einstein. In Arrhenius's general report of 1912, their works were also presented, for the first time, as a *group*. However, as discussed in Chapter 5, by then the "Nernst problem" was probably causing Arrhenius to withdraw his earlier support of Planck and may also have colored his evaluation of Einstein's theory of relativity, dismissed by him summarily as "having not yet gained such a significance that it could be considered as being of greatest benefit for mankind."[70]

Carlheim-Gyllensköld's support of mathematical and theoretical physics did not make him embrace relativity and quantum theory, at least not at this time. After having characterized the nominations of Planck and Einstein as pertaining to "works of a more speculative kind" in the 1914 general report, he noted: "Theoretical physics is undergoing a severe crisis. The principles of the old classical mechanics are contested and new revolutionary opinions seek to gain acceptance, such as the theory of relativity, the atomistic view of energy, etc....on which it is yet too early to pronounce a definitive judgment."[71]

Divergence in chemistry

In contrast to physics, nominators' votes for the chemistry award, especially after 1907, did not offer the chemistry committee much guidance in its choice of prizewinners. Moreover, lacking a common definition of what constituted discoveries of significance for chemistry, committee members experienced greater difficulty in arriving at consensus. This divergence of opinion also characterized the committee's attempts to rely on the statutes to resolve conflict. Considering the range of specialty orientations represented on the committee and the lack of a strong chairman, it could only be a matter of time before the committee experienced serious difficulty in its decision making.

By 1910, awards had been made in organic and inorganic chemistry as well as in the new fields representing the interfaces of chemistry and other disciplines, physics in particular (physical chemistry and radioactivity), but also biology (biochemistry). Ensuring a broad representation of fields thus appears to have been a criterion guiding the committee in its choices. This principle of rotation may also have directed the committee's attention to those fields – chiefly chemical technology – that so far had gone unrewarded. Although perhaps speaking only for himself, Pettersson had signaled such a move when he wrote to a Norwegian colleague in 1905: "In the committee, we have tried so far, first, to reward the great *thinkers* who have provided us with our present chemical theory; subsequently, we want to begin rewarding men of practice, discoverers of useful things."[72]

Not surprisingly, it was Pettersson who took the lead in bringing innovations to the attention of the committee: In both 1907 and 1908, he nominated Swedish inventors (F. A. Kjellin and G. de Laval) for the prize. That the committee came close to awarding a prize for work in metallurgy in 1907 is shown by its decision to have Pettersson draw up a report giving reasons why the prize might be jointly awarded to H. Le Chatelier (professor of metal-lurgy at the Ecole des Mines in Paris) for his theoretical work in this area and to F. A. Kjellin for improvements made in the electric process of cast iron manufacture. When industry representatives in the Academy were consulted about the Kjellin process, how-ever, they expressed reservations.[73] Although this advice caused the committee to drop these particular candidates, it did not aban-don the idea of a prize in metallurgy.

In 1910 P. Martin, inventor of the process bearing his name, was proposed by one of the industry representatives in the Acad-emy who had advised against the award to Kjellin. The Martin process dated back to the 1860s, but it was not until the first decade of the twentieth century that it became widely adopted for steel manufacture. It seemed possible at first that the committee would rally around Martin's candidacy. A special report written by Kla-son stressed that the technical and economic superiority of the process was clearly demonstrated and also that it was high time to reward Martin, who had derived few economic benefits from his work and was now, at the age of eight-five, practically des-titute. A redrafted version of the report, however, ruled out a prize, citing the statutory requirement that the works rewarded be of recent date.[74]

The abandonment of the Martin candidacy was due in part to the prospect that a committee majority could be created for a prize in another area of chemical technology, that of the fixation of atmospheric nitrogen. Several competing methods for nitrogen fixation were then being developed for industrial production in the hope that they would meet rapidly rising needs for nitrogenous fertilizers in particular, but also for the nitric acid that was raw material for the explosives industry. By 1910, the heavy devel-opment costs, most of which had been borne by the German chemical industry (BASF and Siemens, particularly) seemed to be paying off, and the three principal methods for nitrogen fixation had all moved into the production phase. They were (1) the arc process developed by K. Birkeland, professor of physics at the University of Oslo, and S. Eyde, an engineer; (2) a variant of the arc process developed by O. Schönherr; and (3) the cyanamide process developed by A. Frank and N. Caro.[75]

As the main spokesman for an award in this area, Klason initially supported the candidacy of Frank (alone or together with Caro);

but subsequently he focused his attention and, against their will, much of the committee's as well, on Birkeland and Eyde. His arguments, first presented in 1909 when he alone voted for Frank against the committee majority, who supported Ostwald, changed little with time. To supply the world with food required nitrogenous fertilizers; hence, "any invention or improvement in this area is of great benefit to mankind and should therefore be furthered by the Nobel Foundation whose express purpose it is to reward successful enterprises of this kind."[76] The problem, of course, was that the committee could not just recommend "any invention or improvement" for a prize in this area but had to choose the one that presented a definite advantage over previously existing methods and, given the committee's hesitancy to divide the prize, preferably was the work of one individual whose achievements were clearly superior to those of both his colleagues and competitors. On both these counts, Klason presented arguments (too lengthy to be discussed here) designed to overcome the hesitations of his colleagues. In the main, he fought against the opinion that has turned out to be the correct one, namely that these methods would eventually be of historic interest only, as they were likely to be replaced by more efficient and less costly processes.[77] (At the time, the Haber–Bosch process for which Haber won the Nobel prize of 1918 was still in the development stage.)

In 1910, Klason and Söderbaum had both nominated Frank, Klason's nomination proposing, as an alternative, a prize divided between Frank and Caro. In the first voting, Frank's candidacy won a majority of three (Klason, Söderbaum, and Hammarsten), the remaining members each favoring his own candidate. However, the support of Hammarsten, the committee's new chairman, was both tentative and conditional, since he accompanied his vote by a statement in which he warned the committee that "if, *in order to reach agreement*, he were to cast a definitive vote for Frank, he would also feel compelled to present his hesitations concerning this proposal."[78] With the chairman wavering, the weak majority for Frank rapidly dissolved and in the next round there were as many proposals as there were members present; Klason still favored Frank, Widman voted for Grignard (about whose work he presented a special report written on his own initiative), and Hammarsten did not feel that he could cast his vote for any of the year's candidates. The most radical solution to the dilemma, however, was proposed by Söderbaum. He suggested that the committee fall back on para. 5 of the statutes, which states that if no work "proved by the test of experience or by the examination of experts to possess the preeminent excellence...signified by the terms of the Will," the award should be withheld until the following year. In the next and final round of voting, this proposal won, although it only received two votes, that of Hammarsten being the decisive one since one member (Pettersson) was absent. The committee thus decided to recommend to the Academy that the prize for 1910 be reserved.[79]

This solution was not acceptable to the chemistry section of the Academy since, in the words of one of its members, "to everybody interested in the prizes, and to the world of chemistry in particular, it would appear as a mystery and certainly create a stir which could not but reflect badly upon the Academy."[80] Instead, the section directed its efforts toward trying to find a compromise candidate. That this could not be either of the two (Frank and Grignard) on whom the committee had voted was clear for two reasons. First, the claims for Frank as the sole originator of the cyanamide process that Klason had presented to the section were strongly opposed by Hammarsten. Second, although members were not opposed in principle to awarding Grignard the prize for his discovery of the so-called Grignard reagent (which had proven to be of great utility in organic synthesis), a dispute over the priority to this work had recently arisen between Grignard and P. Barbier, his former teacher at the University of Lyon. As Hammarsten pointed out to the section, this was not a good time for rewarding Grignard. Examining the candidates who for one reason or another had been discarded by the committee, the section commissioned a second opinion on Nernst from Arrhenius, who had already presented the committee with a report expressing his doubts about the experimental verification of Nernst's heat theorem. In his report on Nernst's other works, Arrhenius found none of these to be sufficiently recent or important for chemistry to warrant a prize.[81]

After lengthy discussions in the section, a compromise candidate emerged: O. Wallach, proposed for his work on organic compounds belonging to the terpene and camphor groups. Although the committee had earlier bestowed much praise on these "exemplary investigations," which represented "a gigantic piece of work covering over twenty years and yielding more than eighty publications," it had also felt that they were superseded by those of von Baeyer, who had been the first to achieve the synthesis of a terpene. In 1907 and again in 1910, the committee argued that von Baeyer, awarded the chemistry prize of 1905, stood head and shoulders above Wallach not only when it came to work in this specific area but also in his overall contribution to chemistry. Therefore, an award to Wallach could make it appear "as if the Academy had lowered the strict standards for the prizes it had so far felt compelled to maintain."[82] The section was of another opinion, however, for as one of its members stated, if one were to use von Baeyer or Fischer (the 1902 laureate) as the norm, the prize would be awarded only a few times each generation.[83] Reflecting the sentiment that the prize had to be awarded at all costs, the section went through several rounds of voting, which ended in a prize recommendation for Wallach that was subsequently adopted by the Academy meeting in plenary session. In the final voting in the section, most committee members supported Wallach, giving as the reason the need to reach an agreement. Only Pettersson dissented since, in his opinion, if para. 5 was not used sooner or later, both "the worth and the reputation of the prizes would be lowered."[84]

This episode testifies to the importance of committee consensus for the overall process of decision making, since it makes clear that when the committee could not reach an agreement the Academy would step in and force a decision. In this latter event, it mattered little whether the considerations of the committee had been well founded or not, since other concerns, chiefly those of finding a way out of the impasse, would determine the final choice. In the present case, evaluations of leading candidates were few in number and sketchy in content.[85] That the inclination to withhold the prize was caused more by the members' inability to agree than by the lack of merits of different candidates as revealed through evaluations shines through the general report. Hence, it is somewhat ironic that after repeatedly referring to fundamental differences of opinion as the reason for its inability to make a choice, the committee would invoke para. 5 where withholding the prize is made dependent on "the test of experience" and "the examination of experts." In all probability, serious evaluation of the candidates was precluded by the disagreement that already reigned among members at the outset of the year's prize deliberations. As other examples show, lengthy reports on the candidates were most often drawn up at the decision emergence stage where they were both supported by and supportive of the emerging consensus. When there was such strong divergence of opinion, as in the present case, there would be little point in taking the time and effort to draw up such reports. It is interesting to note that the same limitations on the amount of knowledge formalized in special reports obtained when there was strong agreement, although here it was dictated by the lack of necessity for knowledge about the candidates other than the one about whom the members had agreed.

The members' disagreement in 1910 was partly due to the feeling that none of the candidates was worthy of a Nobel prize. This episode also illustrates, then, how the overall standards that had evolved over time influenced the decisions of a given year. When Wallach was compared unfavorably to von Baeyer, for instance, it was not only on the basis of the works for which he had been proposed but rather with reference to the overall reputation he enjoyed in "the world of chemistry," to use a committee report expression. This type of evaluation was largely the result of the practice that had evolved over the first ten years of awarding men who had left their mark on an entire subfield of chemistry. Although the committee had been careful to cite specific discoveries, it was clear that the awarding of van't Hoff, Fischer, von Baeyer, and Ostwald in particular was based on a lifetime's work that was very much out of the ordinary, quantitatively and qualitatively. With this group of great chemists having received awards, it was natural that the committee would experience a certain amount of disenchantment with the new group of candidates that was emerging from the nominations. From this viewpoint, the chemistry section's having forced a decision undoubtedly had the effect of making the committee aware of the necessity both to lower its sights in general and to base its evaluations on specific works, whether or not these constituted "important discoveries."

In the years immediately following 1910, some of the candidates (in particular, Werner and T. W. Richards) whom the committee had previously considered "not deserving" were recommended for the prize.[86] This came about in part through improved conditions for committee decision making, in part through the way nominators' support for these candidacies had been building up over time. With respect to the former, the succession of Hammarsten to the chairmanship undoubtedly made a difference, since he seems to have been more successful than Pettersson in reconciling conflicting opinions. An additional factor was the replacement of Pettersson as a committee member by Å. G. Ekstrand, who seems to have been more diligent and more accommodating than his predecessor. When Klason tried to revive his idea of a prize for nitrogen fixation in 1913, Ekstrand's vote strengthened that of the committee majority and Klason was left as the sole member of the opposition.

Given these new conditions, the committee was better able not only to agree on the choice of a given year but also to plan for upcoming ones. As was already the case for physics, the general reports in chemistry began to be more explicit about the "lineup" of candidates whom the committee intended to recommend for the prizes in the next few years. Planning ahead also allowed the committee to "work off" candidacies in a more expeditious manner, as shown by the prize for Grignard and Sabatier in 1912. The committee justified the divided prize by citing not only the similarities of their methods (both of which had proven to be of great utility in the preparation of organic compounds) but also the consideration that if only one was rewarded, several years would go by before the committee could recommend the other, since both their fields and their nationalities were the same.[87] On the whole, then, it seems that by 1915 the committee had surmounted most of the internal difficulties of decision making. This was largely due to its overcoming several factors impeding consensus, chiefly the absence of leadership from the chairman and the members' inability to work together as a group.

The foregoing has shown that consensus was the key to the physics and chemistry committees' ability to fulfill their mandate. For the Nobel institution to achieve credibility within the context of international physics and chemistry, it was necessary for the committees to dominate the decision making. In spite of internal obstacles, both committees could achieve some degree of stability in their operating procedures that permitted them to evaluate, select, and justify candidates for the prizes relatively independently of the Academy. The committees' creation of their own traditions in these respects was perhaps the most important factor in the long-term functioning of the Nobel institution. The examination of the institution would not be complete, however, without considering its public face – that is, the reception given the prizes by the public and the scientific community at large. This is the subject of the following chapter.

7

The prizes, the public, and the scientific community

From the beginning, the Nobel institution was a subject of intense interest to newspapers all over the world. Initially, this interest concerned the magnificent gift that Nobel had made to promote progress in science and culture; later it switched to the selection of prizewinners, particularly those in literature and peace. In the journals of the scientific professions, however, the prizes did not attract the same interest, nor do they seem to have had the status in the scientific community that they later came to acquire. Acquiring this status would be a slow process, which extended over a longer period than the fifteen years treated here. During the early years, the prizes were nevertheless important in the scientific culture of the time, and this importance also translated into prestige for the institution.

An institution that carries fame and prestige is not necessarily a significant one; for analytical purposes, then, it is important that these two aspects be kept distinct. The fame and prestige of the prizes were being established during the period studied here, rapidly in the case of the public and more slowly within the scientific community. As for the significance of the prizes, however, the views differed and have continued to differ. Then, as now, the science prizes undoubtedly did much to educate the public and the polity about scientific activities, but this would only indirectly influence these activities. A direct influence of the prizes on science in the making was of course not present in the works rewarded in the early years since, in almost all cases, these works dated back to the period before the prizes were instituted. There is no doubt, however, that early on the prizes became significant for the recognition of particular strands or "schools" of research – experimentalists or the Ionists, for instance. In the early years, the prizes were also much more important for the financing of research than they have since become. Their role in this respect came about because the prize money was most often donated to scientific institutions and because the high public visibility of the awards meant that those who had been honored could attract funds to their institutions from government and private sources.

Since intense public interest in the Nobel prizes was manifest from the beginning, and since this interest was to some extent the vehicle through which the prizes became significant for science, it seems appropriate to devote some time to the reception of the prizes in the popular press. Some of the main themes of the publicity surrounding the institution in general and the science prizes in particular will be traced out. An attempt will also be made to assess the extent to which concern with public reactions influenced the choices of prizewinners. In the second section below, the science prizes will be examined in the context of existing national and international reward systems to show how they slowly gained a standing in the international scientific community, by building on these systems and by being more significant for science than other honors.

PUBLIC INTEREST AND ITS CONSEQUENCES

As mentioned in Chapter 3, Nobel's bequest had been reported in newspapers throughout the world. That a solitary millionaire (and the inventor of dynamite, to boot) would leave his entire fortune to promote culture, science, and, what was perhaps most intriguing, *peace* was a story, it was felt, of great interest to the public. Once the awards began, this initial publicity spread in a prodigious manner. By the time the first awards had been announced and commented on (i.e., from December 1901 until mid-1902), the one hundred journals that had reported on Nobel's bequest had grown to around five hundred (not counting those inside Scandinavia). During this period, close to a thousand items were published about or relating to Nobel and his prizes. Of the journals, 34 percent were published in France or its colonies, and these carried 37 percent of the items. Germany accounted for 22 percent of the journals (23 percent of the items), whereas the British Empire and the United States together only represented 12 percent (13 percent of the items). Other countries in different parts of the world (Europe, Latin America, Asia, etc.) accounted for 32 percent of the journals (28 percent of the items).[1] The largest portion of this publicity concerned the prizes in literature and peace, which not only held more interest for the public than the science prizes but which were also the subject of lobbying by different interest groups.

What were some of the reasons for the high degree of interest in France and to a somewhat lesser extent in other countries? In France, initial interest was undoubtedly prompted by the fact that two Frenchmen (Sully Prudhomme and F. Passy) won the first year's prizes in literature and peace, respectively.[2] That this interest also endured was probably due to the prominent role that prizes of all kinds – literary, scientific, and artistic, not forgetting those distributed in the schools – traditionally occupied in French society. As one commentator put it: "In France, we are in school all our lives. . . . It is a need of our race to establish hierarchies, to distribute prizes, titles and awards."[3] Here was a ready-made system, then, on which the press was in the habit of reporting and into which the Nobel prizes were readily assimilated. In other countries, the publicity surrounding the first year's prizes was maintained or grew as each major nation and a few smaller ones built up its stock of prizewinners. One observes, in fact, a "snowball effect," since the press gave not just the formal announcement of the awards and the obligatory interviews with the prizewinners, but also items such as news of the activities of the previous years' laureates, their obituaries, and reports on the creation of other "Nobel prizes." To these should be added, as a particularly important category during the first ten years, endless speculation concerning the presumed prizewinners.

The many rumors that circulated each autumn about who would win the prizes that year were due not only to curiosity, but also to the rule that the names of the prizewinners be kept secret from the public until the award ceremonies, held in Stockholm and Oslo on December 10. That the prize awarders thought it would be possible to conceal the identity of the future laureates, even as they were traveling to the two cities to receive their awards, attests to their having been initially quite unaware of the amount of publicity that would surround the institution. As it turned out, the secrecy presented journalists with a challenge, and they became quite skilled at ferreting out the decisions as soon as they had been made, and sometimes before that. The annoyance that this caused the laureates, who, being sworn to secrecy, had to issue denials of information that they knew to be correct,[4] was nothing compared to the acute embarrassment that ensued when the rumors turned out to be false. One such incident occurred in 1908, when journals all over Europe reported that Planck was to be the recipient of the physics prize (see Chapter 5). This and similar incidents probably prompted the change in procedures instituted in 1910, when the awarding institutions started to announce their decisions once they had been reached. This ended the ritual of speculation and also made for the more controlled publicity that ensued as each prize-winner in turn was presented to the public.

Public interest in the prizes was strongly linked to the view that they were a contest among nations, much like the popular (and equally erroneous) view of the Olympic Games. Given the prevailing climate of nationalism and intense competition among nations, this "contest" was not always presented as a peaceful one. Frequently, achievements that supposedly had benefitted *all* mankind were heralded not just as those of individuals belonging to particular nations – each country's press making the most of its own prizewinners – but as proof of the superiority of that nation in the realm of culture or science. Often the papers went beyond simply priding themselves on the presence of one or more of their own nationals among the year's crop of prizewinners. Commenting on the absence of British laureates in the first year, a French journal wrote: "The British have the insolence to pretend that they are, in all respects, the first nation on this earth. . . . But England has been defeated by the rest of Europe in the peaceful pursuits of the sciences and the arts, that is, in the domains which represent true civilization and real progress."[5] When three of the five prizes went to Germans in 1905, several German newspapers headlined this announcement with the battle cry of Count von Bülow, whose troops had routed the French at Waterloo: "*Deutschland in der Welt voran!*"[6]

As the tensions grew that eventually led to the outbreak of the First World War, the prizes were more frequently viewed from nationalistic perspectives. On the tenth anniversary of the first awards, a German journal inaugurated the statistical computations of the number of prizes won by each country that henceforth would be a stock feature of Nobel prize publicity. Not unexpectedly, Germany came out on top, whereas the performance of the United States, which by then had captured only two out of the sixty-two prizes awarded to date, made the reporter exclaim: "A nation of ninety-three million which only receives one-thirtieth of the prizes!" After pointing out that one of the American prize-winners (Michelson) had in fact been born in Germany, he concluded: "Let us make sure that, in the future, Germany will retain the brilliant position that it has achieved in the first decade."[7] Shortly after the outbreak of war, Ostwald used these same statistics as proof of Germany's superiority, particularly in the sciences. In an interview with a Swedish newspaper widely quoted in the Allied press, he attributed Germany's success to the fact that "we, or rather the Germanic rate, have discovered organization." Whereas the French, British, Russians, and Italians still lived in the era of individualism, the Germans alone had created a culture based on organization.[8] When Arrhenius recommended to the Academy of Sciences in 1915 that the prizes be suspended until the end of the war, he used Ostwald's statements as evidence for the fact that "the belligerent parties have largely lost the ability to treat the Nobel prizes in an objective manner"; hence, "it would be preferable if the prizes were not awarded until more sober judgments could be reached."[9]

That the Nobel prizes came to be used so extensively as the means whereby different nations took stock of their achievements (most often in earnest and not simply for propaganda purposes) not only was an indication that they were taken quite seriously but also required that the decisions receive universal approbation. In this area, as with respect to the sheer volume of publicity already noted, the reaction must have surpassed all expectations. In general, the press did not just approve the prize selections but was also prone to regard the prize awarders as paragons of Solomonic wisdom and historic vision.[10] Commenting on the awards of the first four years, an American journal commented: "Thus far the credentials of all those who have been distinguished by the prizes are approved by the great popular jury as well as the special one that made the award. The list will be historical, and it will help those of today and those who come after to follow the evolution of these higher interests of civilization and humanity."[11] Another journal commented two years later: "The history of modern science might be written without going outside the names of the Nobel prizes for beneficient discoveries in physics, chemistry and medicine."[12] Also indicating the importance almost immediately accorded the institution by the press were the recriminations – in a series of letters addressed to *The Times* of London – when it was discovered that no British men of science or letters had been included among the prizewinners in the first year. In Britain as well as in the United States, it was felt that the absence of a central body to coordinate the proposals of invited nominators constituted a decided disadvantage in competing for Nobel prizes, particularly against France where, it was thought, the highly organized system of academies performed this function.[13]

There is no doubt that the large sum of money the prizes represented was a factor explaining not just why the institution was considered important by the press – wealth in its different forms having always been a subject of fascination – but also why the decisions of the prize awarders were so rarely questioned or challenged. Such enormous sums, it must have been felt, simply *had* to be dispensed in a judicious and prudent manner, *ergo*, the achievements crowned were, without question, superior to all others. That Sweden was not directly in the sphere of interest of one or the other of the Great Powers was another factor explaining the prize awarders' reputation for objectivity. Since the image of Sweden as a politically neutral country did not really take hold until the First World War, what played a role rather was that, on the whole, the prize awarders did not appear to represent any of the dominant systems in culture or science, whether Germanic, French, or Anglo-Saxon. When prizes were awarded to Frenchmen, this took on special significance in the French press because the Swedish prize awarders had been assumed to represent a Germanic tradition. In Germany, it was felt that this factor had not contributed to favored treatment since, after all, the Swedes were often regarded as "the Frenchmen of the North"!

Turning to the popular reception of the prizes in physics and chemistry, there is no doubt that the 1903 physics prize awarded to Becquerel and the two Curies represented a watershed both because of the dramatic increase in the sheer volume of reporting and because of the new forms of publicity about the science prizes that it engendered. The first year's prize to Röntgen had of course been reported, but not extensively so since, by this time, there was something passé about X-rays.[14] When Lorentz and Zeeman shared the physics prize in the following year, only summary accounts of their work appeared in the press. This was probably due to the fact that it was considered too esoteric, especially when compared with the much more popular "discovery" of wireless telegraphy by Marconi, who had been touted in the press as the almost sure winner of the prize in that year.[15] By 1903, then, the prizes in physics and chemistry were far removed from those of literature, peace, and medicine in terms of the response they elicited from the public. This all changed with the award of that year's physics prize to Becquerel and the Curies.

In retrospect, the 1903 physics prize was a masterstroke, since it etched into the public mind an image of scientific work and achievements that considerably enhanced the popular standing of the science prizes. The most important factor here was the Nobel prize for Mme. Curie, which was at the source of her sudden fame and also established her legendary popular status in twentieth-century science.[16] The image propagated was not only linked to the person of Marie Curie, however, for in the future the several themes of which it was composed – the pursuit of pure science for its own sake, the wondrous and totally unexpected results that might flow from such endeavors, and the struggle of the individual to overcome formidable obstacles – were to recur.

In expounding these themes, the press almost without exception seized on one single achievement: the discovery of radium by the Curies and their work to isolate this new element. It is somewhat ironic (though hardly surprising) that this particular achievement was treated as a sensation in the world press, since the physics committee had had to delete the discovery of radium from the award citation so as not to infringe upon the domain of the chemistry prize. Highlighted in varying order of importance in the press were: the magical properties of the new element, which seemed to hold not only an inexhaustible supply of energy but also the capacity to transform itself into other elements; the incredibly high price of even a minute quantity of the substance (the figures quoted being for the most part vastly exaggerated); and its potential utility as a therapeutic agent against cancer, tests (whose results were still inconclusive) having begun in different medical centers in 1902. This wonderful discovery was presented as the result of the incredible persistence the Curies had shown in their experimenting. Here, a poignant image of the experimental scientist at work was forged by the descriptions of Marie Curie grinding down tons of pitchblende in order to extract a minute quantity of radium. The effusive language in which all this was described – even *Le Figaro*, normally a staid journal, told it as a fairytale, complete with "once upon a time" – is well illustrated by a sample from another French journal: "Voilà," exclaimed the reporter, "perpetual motion, the eternal sun, the supreme inexhaustible force have at last been found through the geniuses of the inventors M. et Mme. Curie, whose Nobel prize fits them like hand in glove."[17]

The wonderment that this discovery caused among the public extended to its discoverers. Here, interest was compounded by the fact that the Curies were almost totally unknown to the public. "We do not know our scientists," wrote *La Liberté*, "foreigners have to discover them for us"; and after describing how the Curies had been disregarded by the French scientific establishment, the piece ended with the rhetorical question: "Is that not a good lesson which the luminaries of European science have taught our grantors of official favors and positions?"[18] Whereas for previous laureates reporters had been content simply to reproduce official biographies and accounts of the work honored, they now wanted first-hand knowledge of how the Curies lived and worked. Pilgrimages to the shed in the back of the Ecole Municipal de Physique et de Chimie (the site of the discovery) did not yield much new material, since both Curies showed great reticence in talking to the press about their work, let alone their private lives. It established as an important precedent, however, that scientists could be approached directly to give their views not only on their work, but also on other matters of topical interest. Underlying this was the assumption that the activities and opinions of distinguished scientists interested the public to the same extent as those of other public figures.

Although in the period before the First World War publicity about the science prizes never again reached the same level of intensity as in 1903, the curiosity which that prize had engendered nevertheless had a lasting impact on reporting about the prizewinners and their achievements. Henceforth, reporters and the prizewinners themselves, in the obligatory interviews that followed the announcements, felt compelled to present their work in terms comprehensible to the average reader, whose interest in these matters was no longer in question. Discoveries awarded the prize were expected to involve surprises, startling effects, and leaps

into the unknown. That the atmosphere contained hitherto unknown constituents – the noble gases, or "Nobel" gases, as they
were duly named – that diamonds could be produced artificially
in the electric furnace invented by Moissan, that temperatures
could be lowered to a previously unsurpassed degree as exemplified by the work of Kamerlingh Onnes, to cite only a few examples, were all discoveries *in pure science* that were made to appear
as exciting as the technological advances that were only occasionally rewarded by the prizes. In all these respects, the prizes contributed significantly to the popularization of basic research.

It was generally and correctly assumed that achievements of
such magnitude required long preparation and hard effort – the
image of the experimental scientist toiling away at the bench was
frequently evoked – making for the recurrence of the themes of
self-sacrifice, disinterestedness, and struggle that had been so
prominent in reporting on the Curies. Although scientific work
had previously been presented in this fashion – the legend of Pasteur being the prime example in the nineteenth century – the annual
announcement of the Nobel prizes became the occasion for a recurrent parable in which public attention was focused on a specific
achievement wrought by a single individual. When a latter-day
science writer presents the challenge "Try to imagine physics without Einstein" to his readers, this does not evoke so much the
revolutionary impact of Einstein's work as a powerful image of
the individual *and* his work, which in the case of Einstein, as for
Marie Curie, was highlighted by the publicity surrounding their
Nobel prizes.

This amount of publicity and the forms it took quite naturally
had an impact both on how the prize awarders went about their
tasks and, as will be discussed in the following section, on the
standing of the prizes in national scientific communities. It must
have been apparent from the beginning that the Nobel institution
could neither operate in isolation from the public (for this would
run counter to the educational and enlightenment purposes that it
was hoped the prizes would serve) nor let itself be influenced by
anything that was written about the prizes. That the institution
came to stand for seriousness and high purpose in the public mind
was largely a result of the careful balance that was struck between
the two above-mentioned considerations. The statutory rule that
the prize deliberations be kept secret was the most significant action

in controlling publicity; when originally instituted, this rule had also been justified by the danger of the prizes losing prestige if controversies were aired in public. The most important consequence of this was that once the announcement was made, public attention was narrowly focused on the prizewinners – never on the candidates[19] – and on the prize awarders' own justification for their choice, this being the basis for most of the publicity about the awards. On the whole, the secrecy rule was observed, although during the years when the announcement was not made until December 10, members of the Academy may have leaked the names of the prizewinners to the press. In 1903, following publication by a member of the Academy of his proposal of Marconi, the Academy extended the secrecy rule to nominations submitted by the members themselves or communicated to them.[20] Although foreign nominators were not bound by this rule, it is interesting that, in contrast to the prizes for literature and peace, nominations for the science prizes were hardly ever publicized.

The careful balancing act that produced such vast amounts of "good" publicity about the science prizes – that is, publicity that focused on the prizewinners and therefore generally enhanced the prestige of the institution – was also achieved by the prize awarders themselves staying out of the public view. The secrecy rule precluded anything being said about the process of prize selection, but members of the Nobel committees could still have held forth about general aspects of the institution or matters of topical interest and so emerge as public figures. Given the amount of curiosity, however, anonymity must soon have appeared to be a necessity, since even the most innocuous statement could have been misconstrued or blown out of proportion. This policy was helped by the Nobel Foundation, as the "public face" of the institution, acting as a buffer between the public and the prize awarders. The distance that members of the Nobel Committees for Physics and Chemistry kept from the press was applied to other manifestations of public interest in the prizes. For example, the large number of claims on the prizes received from amateur scientists and inventors all over the world (often exceeding the number of official nominations) were banished to an ever-growing "crackpot file." Acting out of public view and only within the boundaries set by the statutes doubtless made it easier for the committees to perform their task of selecting prizewinners. Still, even considering the geographic distance that separated Sweden from foreign scientific centers, there is something unreal about the almost total absence of references to public interest in the prizes in the letters committee members wrote to one another.

That so little concern was expressed over the popular reception of the award decisions quite naturally makes it difficult to assess the extent to which committee members kept in view public reactions to their decisions and geared their choices to these. While they were not entirely insensible to such reactions, it can fairly be assumed that these were not uppermost in their minds and hardly comparable to the much larger concern shown for the reception of the choices by the scientific community, as illustrated by comments both in letters and in committee reports. The ease with which committee members seem to have been able from the beginning to disregard popular reactions was probably due to their realization that the press, not knowing the names of the candidates, was not in a position to lobby for realistic alternatives. As time went on, the almost universal approbation that met successive prize decisions further increased committee members' propensity to make light of popular reactions. Looking slightly beyond the period studied here, this attitude is exemplified by the comment of a member of the physics committee that, as far as he was concerned, Einstein would never receive the prize for his theory of relativity, this being precluded, among other things, by the intense interest that his theory had aroused among the public.[21]

Still, in a few specific cases the desire to make awards that would elicit favorable popular reactions undoubtedly influenced the decisions. The members of the physics committee – Ångström, in particular – who campaigned so hard for Becquerel and the Curies were probably partly motivated by their feeling that this was a popular choice. Yet they could hardly have foreseen the torrent of publicity it was to produce. Although the discovery of radioactivity in 1896 and of radium in 1898 had not evoked anything like the same enthusiasm as that of X-rays,[22] by 1903 there was no doubt that radium in particular interested a nonspecialist public. This had been demonstrated by the reaction of the large audience who attended Pierre Curie's lecture at the Royal Institution in London in June of that year, by the crowds that gathered at the Sorbonne to hear Marie Curie defend her doctoral dissertation the same month, and, closer to home, by Ångström's own lecture on radium at the Academy of Sciences.[23] As for the inclusion of Marie Curie among the prizewinners, the members of the physics committee were probably aware that, in addition to the justice of her selection, it would also produce what her husband had rather naïvely called an "artistically satisfying" effect. The interest in radioactivity shown by the chemistry committee may also have been fanned by the success of the physicists' initial venture into the field. The rewarding of Rayleigh and Ramsay for their discovery of the inert gases is another instance where the public reaction, anticipated by the interest this discovery had aroused in the scientific community at large, doubtless played a role.[24] Finally, the awards made for work in technology, that to Marconi and Braun in particular, might have been influenced by public criticism that the prize awarders had previously focused too exclusively on discoveries of little practical value and, more specifically, that they had ignored the engineering endeavors that had occupied Nobel himself during a major part of his life.[25]

Concern with popular approval is manifest not only in the award of discoveries with a potential for being popularized but also, more generally, in that of the scientists honored with the prizes. If, in the beginning, this mainly took the form of selecting prominent representatives of national scientific communities whose names would lend prestige to the institution, once laureates began to emerge as public figures other considerations intervened. One of these was the age of the candidates (see Chapter 6); here, concern over the image of a laureate as one of a man in his prime was probably reinforced by the occasional criticism in the press that the prizes had become an old-age pension. Another concern had to do with the laureates' finances, since heaping such large sums of money on wealthy persons, particularly if the discovery or invention for which they were awarded was the source of their wealth, might have led to adverse publicity. This was less of a problem for the majority of academic scientists, whose popular image, even when the facts were to the contrary, was not one of great wealth, than in the case of awards for patented inventions, which, as discussed in Chapter 6, was rendered difficult by the attention paid to financial need. Yet another consideration was the moral and ethical conduct of the prizewinners; in this area Arrhenius seems to have been particularly vigilant. Although his inveighing against Nernst on account of his poor morals was caused more by personal animosity than by a concern for the reputation of the institution, his arguments may still have influenced his colleagues. A more real fear of adverse publicity lay behind his attempt to prevent Mme. Curie from traveling to Stockholm to receive her second Nobel prize in the midst of the uproar in the press over her presumed liaison with P. Langevin, the French physicist. As it turned out, his fears were exaggerated, for when Mme. Curie appeared in Stockholm (after Mittag-Leffler had beseeched her to come in a series of coded messages) there were no adverse reactions.[26]

Although Arrhenius may not have realized it, by this time the image of both the laureates and the institution was basically unassailable, and possible occasional mishaps resulted in no loss of prestige. This was the result neither of committee members assessing what nowadays would be termed the "media impact" of their choices nor of their trying to influence how the prizewinners presented themselves to the public. The way Ostwald lent his prestige as Nobel prizewinner to an endless procession of causes, for instance, must have made any such attempt at control seem futile, even though his activities as a propagandist for Germany during the First World War certainly caused the institution embarrassment. The public image of the prizewinners was forged, rather, by the fact that they were selected from a group of mainly academic scientists in whom the public generally assumed a predominance of high motives – love for their work, disinterestedness in material rewards, and moral integrity. That the majority of prizewinners lived up to this image and often also improved on it – by donating the prize money to their institutions, for instance – quite naturally solidified the reputation both of the Nobel laureates as a group and of the Nobel institution.

THE PRIZES AND THE SCIENTIFIC COMMUNITY

At first, the Nobel prizes enjoyed neither the interest nor the prestige in the international scientific community that they had with the general public. Different types of evidence support this statement.

First, in contrast to the massive buildup in the popular press, the notices that appeared in the leading journals of the scientific profession were both infrequent and scanty. Among the general science journals, *Nature* published only brief résumés of the annual award announcements. In France, the *Revue Générale des Sciences* rather haphazardly did the same for half the years during the period studied here, whereas the *Revue Scientifique* neglected to mention the prizes until 1905, when annual notices began to appear. In the United States, *Science* and *Popular Science Monthly* both ignored the annual award announcement and instead criticized the prize awarders for having violated Nobel's will by diverting funds to local science through the creation of the Nobel Institute for Physical Chemistry.[27] Examination of a sample of specialized disciplinary journals reveals that the prizes only very infrequently received attention.

Second, relatively few of the foreign scientists invited to nominate candidates for the prizes responded, at least in the early years. This caused some concern in the committees, and different schemes for stimulating interest among the nominators were discussed.[28]

Third, an indication of how a small but influential segment of the international scientific community – the prizewinners themselves – viewed the institution is provided by responses to an enquiry launched by *Svenska Dagbladet* on the occasion of the tenth anniversary of the prizes. When asked what, in their opinion, had been the chief role of the prizes to date, the science laureates who answered almost exclusively cited different forms of material advantages: The prizes had enabled their recipients to devote more time to their research and to hire assistants; research laboratories had profited from laureates' donations of the prize money; and so forth.[29] These responses are perhaps even more significant for what they omit. Was the absence of any reference to the prestige of the prizes or to their enhancing the value of scientific discoveries only accidental? Or, as seems more likely, were these and other symbolic aspects of the prizes only slowly coming to the forefront?

The immediacy of the public's reaction and the more gradual one of the international scientific community was largely due to differences in perspectives and expectations. Public interest embraced all the prizes and was mostly determined by the ease with which the achievements honored in literature and peace could be grasped. Scientists, by contrast, paid the most attention to choices in their own disciplines, where their specialized knowledge made them regard these with a great deal of circumspection. It was also natural that they would judge the Nobel institution by reference to the reward systems with which they were familiar, that is, the honors (prizes, medals, etc.) dispensed by national scientific societies. Those acquainted with these systems were well aware how difficult it was to reward individual scientific achievement without being too much influenced by institutional, political, or interest-group considerations. These difficulties must greatly increase, it was felt, when it came to administering the first truly international prize, especially since this involved such large sums of money.[30]

In addition, some crucial provisions in Nobel's will and the statutes may have seemed questionable, and there may have been some apprehension about how the Swedish prize awarders, the majority of whom were unknown internationally, would interpret them. The consternation shown by both nominators and commentators with respect to the requirement that the discoveries honored be of recent date may have been due to the fact that, had this provision been stringently applied, it would have represented a break with the tradition of having major honors crown achievements carried out over a much longer period. Similarly, the provision that the discoveries rewarded be those having conferred greatest benefit on mankind suggested that research with proven practical applications, rather than the basic research achievements highlighted by existing reward systems, would come to the front.

That the scientists chosen for the Nobel prizes were primarily those who had already received the major honors of national reward systems undoubtedly alleviated these fears. It also hastened recognition both because such honors had endowed their recipients with prestige and authority that carried over to the new institution and because selecting such figures provided a guarantee that the choices would meet with the approval of national scientific elites. For example, during this period all the French scientists receiving Nobel prizes in physics or chemistry had previously been awarded prizes or medals by the French Academy of Sciences, these being in most cases such major ones as the *prix Lacaze* or the *prix Jecker*. Likewise, with only two exceptions (W. H. and W. L. Bragg), all the British prizewinners had already figured among the recipients of one or another of the three most important medals of the Royal Society of London (the Copley, the Rumford, and the Davy). Most important, prizewinners were often chosen from the small group of scientists who had received the major honors that these societies regularly bestowed on foreigners and that were recognized internationally as carrying prestige. The high proportion of Rumford and Davy medalists among the prizewinners, for instance, is an indication of how committee members took their cues from highly regarded "international" honors.

By mainly rewarding scientists who had already accumulated significant national and international honors, the Swedish prize awarders made it known to the scientific community at large that the new prize should, indeed, be ranked among the highest honors that "the world of science" could bestow on one of its own. While some time would elapse before this became so, at least one attribute of high rank – the long waiting period that preceded the awarding of particular candidates – early became the expected norm for the prizes. This is illustrated by the comment of Rutherford (who won his prize after he had received the Rumford medal but not yet the Copley) that he had not expected to "be considered as a candidate for several years to come."[31]

One can understand better how the prizes came to take on symbolic significance within the international scientific community if they are placed in the context of the complex relationship that characterized internationalism and nationalism in science. There is no doubt that, when the institution was founded, internationalism had evolved from a commitment to abstract values or mere rhetoric to the very real experience of international collaboration that scientists in many countries had gained through meetings, exchange of publications, and collaborative work. As Schroeder-Gudehus has stated: "The norms of science had an international constituency. If any proof were needed, it is certainly provided by the institution of the Nobel prizes: the very possibility of creating the award depended entirely on the worldwide recognition and acceptance of the criteria the awarding body would bring to bear on its decisions."[32] The extent to which this possibility was realized was due to the institution meeting several of the needs that the new internationalism in science had engendered.

First and foremost, there was a need for a supranational arbiter of the prodigious amount of research results that were being circulated internationally. Here, the Nobel institution filled a void because of the high public visibility of the prizes and because the committees made some effort to avoid having the prizes dominated by any one nation. To gain the attention of a large number of scientists, it was not necessary for the awards to bear upon the science in the making about which national and international specialty groups continuously passed judgment. Rather, to become symbols of the general worthiness and common purpose of international scientific collaboration, it was sufficient that they highlight the work of internationally known scientists and, more important, that the international character of the prizes be upheld.

Second, the complex relationship of internationalism and nationalism in science is most apparent, and also most difficult to demonstrate, on the level of scientists' values; here the belief in the worth and utility of international scientific collaboration often coexisted, despite apparent contradictions, with a commitment to national interests, particularly in the way these were linked to science. The Nobel prizes did not of course resolve the tensions between these two sets of values, tensions that were to boil over during and after the First World War, but they met a need by providing, for the time being, some assurance that nationalism and internationalism in science were not incompatible. For example, in reacting to *specific award decisions*, members of national scientific communities could share the pride of the press and the public when their compatriots were honored,[33] while at the same time the *institution as a whole* was regarded as reflecting mainly international values. Viewing the prizes as a "peaceful contest" among nations represented a similar smoothing over of potential tension between nationalist and internationalist values.[34]

Third, as Schroeder-Gudehus has also suggested, international scientific collaboration was in many respects a logical outgrowth of what was known as *Grossbetrieb* in science – that is, national scientific enterprises such as the German one, dominated by large research institutes, division of labor, and heavy investments in material and equipment. But these trends also engendered reactions, some of the manifestations of which could be observed on the international level. Many scientists turned their attention to international organizational efforts because these seemed to offer relief from the tensions inherent in *Grossbetrieb*. For example, those who feared that national academies were losing influence to other bodies (learned societies or government departments) entertained the hope that the International Association of Academies (IAA), founded in 1899, would exert the kind of authority in international scientific collaboration that, by reflection, would reaffirm the position of the academies on the national level.[35] It is probably no accident that many of those involved with the IAA (for example, Hale, Schuster, and Darboux) were also active as nominators for the Nobel prizes because these latter, too, could be seen as adding an international dimension, and hence new authority, not just to the Swedish Academy of Sciences but to academies in general.

More important, the prizes took on symbolic significance because they revived, on the international level, an emphasis on individual achievement that seemed to be receding from the day-to-day experience of national scientific communities. Scientists who deplored the deemphasis of the creative efforts of the individual that they felt was the consequence of *Grossbetrieb* and feared that they would be reduced to an anonymous mass of research workers could take heart in the way individual achievement was singled out for reward by the Nobel prizes. Here they may have looked to an international prize because national reward systems had had to adjust to the realities of organized research. Being more differentiated in the sense of constituting a hierarchy of awards, these latter systems could provide for the rewarding of both senior and junior collaborators in important research collaborations, whereas the statutory provisions for the Nobel prizes rendered this difficult. The acquiescence of the scientific community to the norm that only the major figure in a research collaboration would receive the Nobel prize[36] may be explained, then, by the fact that collaborators were often rewarded on the national level; and because this was so, the largest international award could gain symbolic significance by reassuring scientists that there was still room for the creative genius of the individual.

The several ways in which the Nobel prizes took on significance for a wider audience of scientists in different countries certainly enhanced the reputation and prestige of the new prizes. For the institution to grow into an important one, however, required not only that it carry symbolic significance but also that it appear to have a bearing on ongoing scientific activities. From the start, the Nobel institution was marked off from existing reward systems in two important respects. The institution added more significant resources to national scientific enterprises and provided these with new means to procure funds from other sources. In addition, the recognition that the prizes bestowed on prominent representatives of different specialties and research areas heightened the visibility of these latter. In this manner, the institution contributed to the efforts of organization and legitimation that are essential in order to maintain existing fields and, more important, to develop new ones. By perceiving the institution as useful in this respect, the specialty communities that gained prominence from the prizes came to form constituencies that slowly extended the influence of the institution in the scientific community at large.

The utility of the prizes for the financing of research came about mostly because they were awarded to established scientists (chair-holders and directors of research institutes) who could capitalize on them in a number of different ways. Awarded in this manner, the prizes became more useful than they would have been if Nobel's idea that they function as a form of financial assistance for younger scientists had been realized. The most direct way in which the prizes provided financing for research was through laureates using some or all of the money for their own work or that of their research laboratories. Here, Nobel prizewinners continued the custom of many scientists of regarding prize money as a form of subsidy for research (see Chapter 1).

The difference in size between existing prizes and the Nobel prizes, however, was highly significant: In France, for instance, one Nobel prize was equivalent to *all* the money paid out annually from the prize funds of the French Academy of Sciences in the period 1901–1910. No doubt the Nobel prizes were seen as more important research subsidies (and probably also more extensively used as such) by British and French laureates than by German ones, who received more support both from government and from private industry.[37] The British prizewinners interviewed in the *Svenska Dagbladet* survey mentioned earlier were also the most insistent on the role of the prizes in financing research; Ramsay, for example, stated that "in England, where private benefactions have to replace government support, the prize money makes it easier to procure apparatus and assistance."[38] By contrast, when the prizes were criticized in a German scientific journal, it was partly on the grounds that none of the German laureates had been lacking in research funds either before or after receiving the prize.[39]

The broader utility of the prizes for financing research arose from their being used as arguments for increasing the funds for the institutions or fields where the laureates were working as well as for scientific research in general. This was done with much more success than for previously existing prizes because of the importance that the press attached to Nobel prizes as a matter of *national* interest and concern. The propaganda value of the prizes was probably most effectively exploited in France, where members of the nation's science establishment did not hesitate to make public the plight of prominent scientists. This was apparent in the reporting on the Nobel prize for the Curies, whose scandalously poor working conditions were cited in arguing that the government should provide support for these two who had brought honor to France. In the event, a chair was created at the Sorbonne in 1904 specifically for Pierre Curie, but further pressure had to be applied in order to procure adequate research facilities for both Curies.[40] This success notwithstanding, the plight of the Curies continued to figure prominently in press campaigns launched to increase the government's support of scientific research: Here, the Curies' *baraquement* (shed) was used to illustrate the general lack of resources for scientific research, just as had been done a generation earlier with Pasteur's *grenier* (attic).[41]

The propaganda effect of the Nobel prize in chemistry jointly awarded Sabatier and Grignard in 1912 closely paralleled that of the Curies'. For Sabatier, who had spent his entire career at the University of Toulouse, the Nobel prize had the effect of attracting new funds from different sources (the state, the municipality, and private industry) to the research institutes, especially the Institute of Chemistry, which he very actively sought to develop. In the ongoing debate over the need to decentralize scientific research, bringing it closer to regional interests, the fact that both prizewinners had spent their careers at provincial universities – Grignard held the chair of organic chemistry at Nancy – was invoked both to show the worthwhileness of government efforts to improve research facilities in the provinces and to stimulate further action.[42] These examples refer only to the benefits that accrued to the institutions of the prizewinners in the immediate aftermath of the awards. As major research centers (universities as well as the specialized institutes associated with these) began to accumulate prizes, these benefits would naturally grow and become more varied as a result of the aura of success that came to surround particular institutions.[43]

The role that the prizes assumed for the recognition and legitimation of particular specialties and research areas represented another aspect of the popularization of the discoveries honored by the prizes. Broadly, the popularizing roles of the prizes involved the general, lay public, but they can also be more narrowly construed so as to concern mainly other scientists. In this latter respect, they were linked to the efforts undertaken, particularly by those pioneering a new field, to convince other scientists of the validity and worth of their contributions, to lay claim to resources (both funds and academic positions), and, more generally, to promote the intellectual and social cohesiveness of the field. These functions of the prizes come into view particularly with respect to the new interdisciplinary research areas that were highlighted in the early prize decisions; they would become more apparent after the First World War, when specialization and the creation of new fields would increase at a much more rapid rate. The utility of the prizes for the recognition of new fields should not make one forget, however, that in the early years scientists mainly perceived the prizes as being useful for the maintenance of existing, well-established orientations, among them experimental physics. The reply of the British experimentalists to Mittag-Leffler's entreaties of support for Poincaré, for example, clearly showed a proprietary interest in the Nobel institution as one that represented the experimental side of physics and that furthermore had to be defended against incursions from other fields.

A comparison of what specialists in modern physical chemistry, on the one hand, and radioactivity studies, on the other, saw as the implications of the awards that directly concerned their fields brings out some of the specific utility of the prizes and shows how this varied from one field to another depending on their respective states of development.

At the time those who had pioneered modern physical chemistry received their awards, cognitive developments in the field were not as important as they had been in the 1890s, when the most productive topics that had arisen in the debates over the physical theory of solutions were being exploited.[44] Hence, to observers of contemporary German physical chemistry, the chief utility of the awards lay in their promoting the institutionalization of the field by helping to secure the appointment of the second generation of Ionists to chairs at German universities. For example, at the time of the award to van't Hoff, the *ordinarius* professorship in inorganic chemistry in Göttingen was being filled; but in the opinion of R. Abegg, who had been one of Arrhenius's postgraduate students in the 1890s, the situation looked unfavorable to "us Ionists." "Therefore, I am particularly happy," he wrote to Arrhenius, "that van't Hoff received the Nobel prize since it represents the approbation of this branch of chemistry by impartial people."[45] That the awards to the Ionists were made in chemistry rather than in physics (as Arrhenius had wished in his own case) was also important for at this stage the success of the specialty had to be achieved through the appointment of physical chemists to positions in departments of inorganic chemistry, where they often met with opposition from the chairholders, rather than (as had been done in the pioneering phase) through the creation of special chairs.[46]

By contrast, in radioactivity research the main utility of the awards lay in their drawing attention to cognitive developments, both the initial discovery of radioactive substances and the explanations for radioactive change provided by the Rutherford–Soddy transformation theory. Although this theory encountered surprisingly little opposition when it was first propounded in 1902–1903, even by 1908 it was still very much a working theory in the sense that its productive value in suggesting new lines of research – in physics as well as in chemistry – had only begun to be exploited. Here, the prize awarded to Rutherford was particularly important because it confirmed the standing of the theory as the predominant paradigm in the field.

The several awards made in radioactivity undoubtedly also helped the organizing efforts to which Rutherford, in particular, was devoting much of his time: creating interest in the new field through lectures and attendance at meetings, recruiting researchers and students into the field, and establishing lines of communication, especially internationally.[47] At this early, organizing stage, it mattered less than it would subsequently that the awards did not distinguish between the physics and chemistry components of the field; in fact, their very ambiguity in this respect may have heightened their utility since early workers in the field had to secure support and resources wherever they could be found, that is, from physicists as well as from chemists. The openness of Rutherford with respect to his own award was probably not atypical of this attitude; after the initial surprise had worn off, his transformation into a chemist through the Nobel prize "remained to the end a good joke against him which he thoroughly appreciated."[48] More important, his reward probably increased the awareness among chemists of the importance of radiochemistry for their discipline and may also have made them more willing to accept the fact that the initial contributions to this area had been made by physicists.[49]

Outside the specialist communities that were growing up around new fields, the Nobel institution acquired wider respect among scientists mainly through the way the prize decisions reflected mainstream concerns in both physics and chemistry. In general, the members of these disciplines agreed on the worth of the experimentalist orientations that were given such strong emphasis in the decisions. Hence, the succession of awards celebrating the major discoveries of late-nineteenth-century experimental science elicited much the same response from scientists as they did from the general public. In important respects, then, the popularizing role of the prizes among the public overlapped with the one they performed among scientists.

Still, there were differences, since the scientists naturally tended to regard the awarding of major discoveries in the specific context of developments in their disciplines. For those physicists who partook of the *fin-de-siècle* mood that held sway in the discipline to a greater or lesser extent (historians of science differ in their viewpoints[50]), the rewarding of these discoveries probably rekindled the enthusiasms they had evoked when they were first announced. The interpretations of the initial discovery that had been advanced in the meantime also meant that their significance had been more fully understood, a fact that could not fail to enhance the value of the awards and hence of the institution. The value of both was also safeguarded because the discoveries honored did not throw serious doubt on either the utility of the experimental approach or the validity of general physical principles. For the majority of physicists, then, the awards probably confirmed Planck's assessment in 1914 that "never has experimental physical investigation experienced so strenuous an advance as during the last generation, and never probably has the perception of its significance for human culture penetrated into wider circles than it does today."[51]

Epilogue and Conclusions

When the First World War broke out, the Nobel Committees for Physics and Chemistry were about to meet to decide on their recommendations for the prizes of 1914. At first, it did not seem as if the war would interfere with the process of prize selection. The committees made their choices – von Laue for the prize in physics and T. W. Richards for that in chemistry – and these were also endorsed by the Academy sections. The recommendations did not reach the Academy's plenary meeting, however, for meanwhile the Nobel Foundation, acting on behalf of the Swedish prize awarders, had successfully petitioned the government to be allowed to defer the prize decisions until 1915.[1] The government's permission was necessary in view of the fact that the statutes contained no reference to *force majeure*, whether it be war, revolution, or natural disaster, as a reason for not awarding the prizes. The only legitimate way that the prize awarders could reserve the prize was by deciding that no prizeworthy works were at hand (para. 5 of the statutes).

The hopes that the war would soon be over had not been fulfilled in 1915; on the contrary, the conflict had been extended. By then, many scientists were not only in active duty in the military but also participating in the propaganda efforts whereby each side sought to depict the other as the unrelenting enemy.[2] These circumstances were invoked by Arrhenius when he sought to convince the Academy, in November 1915, that the science prizes should not be awarded until after the end of the hostilities. He argued that Ostwald, in particular, was using the Nobel prizes for propaganda purposes "in a manner which can hardly be agreeable to us" and that the same could be said of "the Britisher Ramsay, not to mention Kipling." An even more compelling reason was the fact that many of the candidates were performing combat duty and that some, like H. G. J. Moseley, had already died in battle. There was a danger, then, that the men to whom the Academy had decided to award the prize would no longer be alive at the time of the prize ceremony. If this were to happen, Arrhenius concluded, it would not only "create a scandal" but it could well be that the ceremony would amount to a "funeral rite."[3]

Arrhenius's arguments show only limited understanding of the extent of the disaster that had befallen Europe and European science; therefore, it is perhaps not surprising that the majority of Academy members decided that, rather than asking the government for a postponement until the end of the war, they would only put in for another year's extension. When the government was petitioned to this effect, however, it refused to grant the extension. Given this refusal, the Academy saw no other way than to proceed with the prize decisions that came to involve the awards for both 1914 and 1915.[4] The Karolinska Institute took a bolder course of action, however, deciding to reserve its prize by invoking para. 5 of the statutes. This had been done by the Norwegian Storting already in 1914: a logical decision, since the outbreak of war could be considered as sufficient proof of the absence of prize-worthy works in the area of peace. Although this logic did not apply to the prizes in science and medicine, the Academy nevertheless followed the example of the Karolinska and did not award any prizes until the end of the war. During this period, the continuity of the prize selections was ensured because each year both the Academy and the Karolinska invited nominations and evaluated them.

The decision of the government and the prize awarders to maintain continuity without visibility throughout the war shows that, in a relatively short time, the prizes had become an important feature of official life in Sweden and even of the country's foreign policy. Only twenty years earlier, Lindhagen and Sohlman had turned to the government for help in implementing Nobel's wishes. Then it had been a question of making possible the creation of the institution by presenting the Nobel prizes as a matter of national concern. Now not only had this been accomplished but the prizes had also secured a place for Sweden in the world of science and culture. Henceforth, a decision as serious as whether or not the prizes should continue to be awarded in time of war had to be taken in light of political considerations. Here, the fear that the unconditional neutrality that Sweden had declared at the outbreak of the war might in any way be compromised by the prize decisions or by the laureates being used, as they were sure to have been, in the propaganda of the belligerent parties certainly played a role in the decision not to award the prizes during the war. Such a politicization was the price that had to be paid for success, but somehow it was linked more to the intense publicity that had surrounded the prizes from the beginning than to what had really been accomplished during the first fifteen years. That these accomplishments were substantial ones becomes apparent when one considers their implications both for Swedish science and for the awarding of the prizes in the future.

That the executors and their attorney had been able to overcome both the opposition of Nobel's relatives and the vacillations of the future prize awarders was a major accomplishment without which Nobel's intentions could never have been realized. To do so it had been necessary to go beyond these intentions and to introduce incentives – primarily the Nobel Institutes – that would make the task of awarding the prizes appear more attractive because it would benefit the institutions that Nobel had selected for this task. But the strategy employed in the battle over the will had also included a larger vision of the Nobel prize institution as a means of internationalizing Swedish science by promoting collaborations between Swedish and foreign scientists within the framework of the Nobel Institutes. This was not realized in the manner in which it had been envisioned since the only Nobel Institute in the sciences created before the First World War – Arrhenius's Institute for Physical Chemistry – was both too small and too personal to perform this role. Here the absence of a "critical mass" of researchers in physics and chemistry – the weakness that Pettersson and Arrhenius had hoped to correct through a big Nobel Institute – undoubtedly played a role, as did the related circumstance that, in Sweden, these two disciplines were still overshadowed by natural history (in its traditional and modern forms) when the Nobel prizes were instituted.

Notwithstanding the abandonment of this larger vision, the prizes themselves became important for promoting contact between Swedish and foreign scientists. Each year's proposals provided the members of the Nobel committees with an overview of major achievements in the international disciplines of physics and chemistry, which could not have failed to expand their horizons. The annual prize ceremony also became the occasion for many members of the Swedish scientific community to meet with prominent foreign scientists and to listen to the personal accounts of their award-winning work presented in the Nobel lectures. Also, there is no doubt that the prizes facilitated the entry of Swedish scientists into research laboratories abroad: those of the laureates but, perhaps more importantly and also more easy to understand, those of the candidates. Eventually, such contacts would bring about the collaborative research projects that had been part of the initial idea of the Nobel Institute. That the effects of the prizes in some of these respects were apparent very early is shown by the impression of Bjerknes, as an "outside" observer, that "Stockholm will develop into a scientific center of high order. A couple of the most prominent representatives of physics and chemistry will journey here each year and soon we shall also see the creation of model laboratories."[5]

However, the undertaking of such a new and important task as that of awarding the Nobel prizes also created tensions within the small community of scientists in Sweden; in fact, it would have been surprising if this had not occurred. Some scientists were doubtless jealous of the benefits – material as well as in terms of power and authority – that accrued to the small group of physicists and chemists who served on the Nobel committees. Others felt that the prizes introduced elements alien to the normal pursuit of scientific activities. Only a few years after Bjerknes had lauded the effects of the prizes on Swedish science, he wrote to his wife that he was glad to return to Norway since this would spare him further contact with "Nobel corruption."[6] Bjerknes can hardly have used the term literally; instead, it probably reflected his disappointment that he had not been elected to the Nobel Committee for Physics mainly because of his Norwegian nationality and his disapproval of the fact that such considerations had influenced the election. It should also be seen in the context of the abandonment of the hope that the Nobel Institute would have an effect on the general conditions of scientific work in Sweden, for by the time Bjerknes wrote his letter it had become clear that when an institute was set up, it would mainly benefit Arrhenius.

Disappointments such as these do not diminish the accomplishments of the prize awarders, who on the whole acquitted themselves of their difficult task in an honorable manner. That the Nobel prize was the largest prize ever created and that it had been understood from the beginning to be international made these men set their sights higher than they would have had the awards merely represented an addition to a hierarchy of prizes such as existed in the French Academy of Sciences. These high aims reflected the sense of historic mission that had been such an important part of the strategy that Sohlman and Lindhagen had employed to forge the different prizes and prize-awarding institutions into a common enterprise. Here, the creation of structures that separated the function of awarding the prizes from those normally performed by these institutions had the effect of enhancing the standing of the prizes, thus making it easier to realize these aims. To make the system of prize selection operative, however, it was necessary for the Nobel committees both at the Royal Academy of Sciences and at the Karolinska Institute to assume a pivotal role; that the basic decision-making authority rested in a small group not only facilitated agreement on the choices but also made it possible to build up traditions that could guide future choices.

Although the statutes did not set specific standards for the awarding of the prizes, the fact that the prize awarders were bound by common rules nevertheless introduced an element of emulation that protected the institution as a whole against a decline in standards. Competition over the most prizeworthy works was one aspect of such an emulation: Quite naturally, it was most strongly felt in the fields that straddled two prize domains. Such competition operated not only, as has been shown here, between the Nobel Committees for Physics and Chemistry in the area of radioactivity but also between the chemistry committee and the physiology or medicine committee in the area of physiological chemistry. Another aspect of the emulation that existed between the committees was a tacit agreement that the prestige of the prizes not be compromised by decisions that intervened in ongoing debates over the validity of a particular discovery or theory. While this made for generally conservative choices, it created a sense of trust in the prize awarders that facilitated the recognition of the prizes in the scientific community at large.

A third aspect of the way in which the mutual patterning of committee actions affected the overall enterprise concerned the idea that the awards highlight developments in particular specialties and research areas. In the past, major scientific honors (the Copley and Rumford medals or the *prix Lacaze*) had mainly celebrated the achievements of individual scientists, generally the most successful ones in each successive generation. While this was true for the Nobel prizes as well, there was also the ambition to make consecutive awards capture what had been significant overall advances in given fields of science. The string of awards made in "ray" physics or, looking beyond physics and chemistry, the preeminence of bacteriology in the prize selections in physiology or medicine[7] both testify to the realization of this ambition. It was due as much to the committees' giving the term *most important discoveries* a meaning that took into account their role in opening up new fields as it was to the somewhat fortuitous circumstance that works in fields such as ray physics and bacteriology, which had experienced rapid growth and significant breakthroughs in the late nineteenth century, were ripe for award when the Nobel prizes were instituted. While making the awards reflect the developments that had shaped these and other new fields facilitated both decision making about the prizes and their recognition in the early period, it did not necessarily smooth the path for the awarding of the prizes in the future. This would become apparent after the First World War when the physics committee started to grapple with the awards in theoretical physics (those of Planck and Einstein, in particular), which could no longer be put in abeyance since doing so would have meant abandoning the notion that the prizes should reflect significant developments in the disciplines concerned. Here, the hiatus in the prize awarding caused by the war provided the committee with a pause for reflection and an opportunity to start afresh in the postwar period.

Provisions in the will of Alfred Nobel establishing the prizes

The whole of my remaining realizable estate shall be dealt with in the following way: the capital, invested in safe securities by my executors, shall constitute a fund, the interest on which shall be annually distributed in the form of prizes to those who, during the preceding year, shall have conferred the greatest benefit on mankind. The said interest shall be divided into five equal parts, which shall be apportioned as follows: one part to the person who shall have made the most important discovery or invention within the field of physics; one part to the person who shall have made the most important chemical discovery or improvement; one part to the person who shall have made the most important discovery within the domain of physiology or medicine; one part to the person who shall have produced in the field of literature the most outstanding work of an idealistic tendency; and one part to the person who shall have done the most or the best work for fraternity between nations, for the abolition or reduction of standing armies and for the holding and promotion of peace congresses. The prizes for physics and chemistry shall be awarded by the Swedish Academy of Sciences; that for physiological or medical works by the Caroline Institute in Stockholm; that for literature by the Academy in Stockholm, and that for champions of peace by a committee of five persons to be elected by the Norwegian Storting. It is my express wish that in awarding the prizes no consideration whatever shall be given to the nationality of the candidates, but that the most worthy shall receive the prize, whether he be a Scandinavian or not.

The statutes of the Nobel Foundation
and the special regulations concerning
the distribution of prizes from the Nobel
Foundation by the Royal Swedish
Academy of Sciences

CODE OF STATUTES

OF THE

NOBEL FOUNDATION.

GIVEN AT THE PALACE IN STOCKHOLM,
ON THE 29TH DAY OF JUNE IN THE YEAR 1900.

OBJECTS OF THE FOUNDATION.

§ 1.

The NOBEL Foundation is based upon the last Will and Testament of Dr. ALFRED BERNHARD NOBEL, Engineer, which was drawn up on the 27th day of November 1895. The paragraph of the Will bearing upon this topic is worded thus:

"With the residue of my convertible estate I hereby direct my Executors to proceed as follows: They shall convert my said residue of property into money, which they shall then invest in safe securities; the capital thus secured shall constitute a fund, the interest accruing from which shall be annually awarded in prizes to those persons who shall have contributed most materially to benefit mankind during the year immediately preceding. The said interest shall be divided into five equal amounts, to be apportioned as follows: one share to the person who shall have made the most important discovery or invention in the domain of Physics; one share to the person who shall have made the most important Chemical discovery or improvement; one share to the person who shall have made the most important discovery in the domain of Physiology or Medicine; one share to the person who shall have produced in the field of Literature the most distinguished work of an idealistic tendency; and, finally, one share to the person who shall have most or best promoted the Fraternity of Nations and the Abolishment or Diminution of Standing Armies and the Formation and Increase of Peace-Congresses. The prizes for Physics and Chemistry shall be awarded by the Swedish Academy of Science (Svenska Vetenskapsakademien) in Stockholm; the one for Physiology or Medicine by the Caroline Medical Institute (Karolinska institutet) in Stockholm; the prize for Literature by the Academy in Stockholm (i. e. Svenska Akademien) and that for Peace by a Committee of five persons to be elected by the Norwegian Storting. I declare it to be my express desire that, in the awarding of prizes, no consideration whatever be paid to the nationality of

Note: The official translation of the statutes and special regulations originally made in 1900 has been retained even though it is defective in many ways.

the candidates, that is to say, that the most deserving be awarded the prize, whether of Scandinavian origin or not."

The instructions of the Will as above set forth shall serve as a criterion for the administration of the Foundation, in conjunction with the elucidations and further stipulations contained in this Code and also in a deed of adjustment of interests amicably entered into with certain of the testator's heirs on the 5th day of June 1898, wherein subsequent upon the arriving at an agreement with reference to a minor portion of the property left by Dr. NOBEL, they do affirm and declare, that : "By these presents we do acknowledge and accept Dr. NOBEL'S Will, and entirely and under all circumstances relinquish every claim for ourselves and our posterity to the late Dr. NOBEL'S remaining property, and to all participation in the administration of the same, and also to the possession of any right on our part to urge any criticism upon the elucidations of, or additions to, the said Will, or upon any other prescriptions with regard to the carrying out of the Will or the uses to which the means accruing from the bequest are put, which may either now or at some future time be imposed for observance by the Crown or by those who are thereto entitled;

Subject, nevertheless, to the following express provisoes:

a) That the Code of Statutes which is to serve in common as a guide for all the corporations appointed to award prizes, and is to determine the manner and the conditions of the distribution of prizes appointed in the said Will, shall be drawn up in consultation with a representative nominated by Robert Nobel's family, and shall be submitted to the consideration of the King;

b) That deviations from the following leading principles shall not occur, viz.:

That each of the annual prizes founded by the said Will shall be awarded at least once during each ensuing five-year period, the first of the periods to run from and with the year next following that in which the Nobel-Foundation comes into force, and

That every amount so distributed in prizes in each section shall under no consideration be less than sixty (60) per cent of that portion of the annual interest that shall be available for the award, nor shall the amount be apportioned to more than a maximum of three (3) prizes."

§ 2.

By the "Academy in Stockholm", as mentioned in the Will, is understood the Swedish Academy — Svenska Akademien.

The term "Literature", used in the Will, shall be understood to embrace not only works falling under the category of Polite Literature, but also other writings which may claim to possess literary value by reason of their form or their mode of exposition.

The proviso in the Will to the effect that for the prize-competition only such works or inventions shall be eligible as have appeared "during the preceding year", is to be so understood, that a work or an invention for which a reward under the terms of the Will is contemplated, shall set forth the most modern results of work being done in that of the departments, as defined in the Will, to which it belongs; works or inventions of older standing to be taken into consideration only in case their importance have not previously been demonstrated.

§ 3.

Every written work, to qualify for a prize, shall have appeared in print.

§ 4.

The amount allotted to one prize may be divided equally between two works submitted, should each of such works be deemed to merit a prize.

In cases where two or more persons shall have executed a work in conjunction, and that work be awarded a prize, such prize shall be presented to them jointly.

The work of any person since deceased cannot be submitted for award; should, however, the death of the individual in question have occurred subsequent to a recommendation having been made in due course for his work to receive a prize, such prize may be awarded.

It shall fall to the lot of each corporation entitled to adjudicate prizes, to determine whether the prize or prizes they have to award might likewise be granted to some institution or society.

§ 5.

No work shall have a prize awarded to it unless it have been proved by the test of experience or by the examination of experts to possess the preeminent excellence that is manifestly signified by the terms of the Will.

If it be deemed that not one of the works under examination attains to the standard of excellence above referred to, the sum allotted for the prize or prizes shall be withheld until the ensuing year. Should it even then be found impossible, on the same grounds, to make any award, the amount in question shall be added to the main fund, unless three fourths of those engaged in making the award determine that it shall be set aside to form a special fund for that one of the five sections, as defined by the Will, for which the amount was originally intended. The proceeds of any and every such fund may be employed, subject to the approval of the adjudicators, to promote the objects which the testator ultimately had in view in making his bequest, in other ways than by means of prizes.

Every special fund shall be administered in conjunction with the main fund.

§ 6.

For each of the four sections in which a *Swedish* corporation is charged with adjudicating the prizes, that corporation shall appoint a Committee — their Nobel-Committee — of three to five members, to make suggestions with reference to the award. The preliminary investigation necessary for the awarding of prizes in the Peace-section shall be conducted by the Committee of the Norwegian Storting, as laid down in the Will.

To be qualified for election on a Nobel-Committee it is not essential either to be a Swedish subject or to be a member of the corporation that has to make the award. On the Norwegian Committee persons of other nationalities than Norwegian may have seats.

Members of a Nobel-Committee may receive reasonable compensation for the labour devolving upon them as such, the amount to be determined by the corporation that appoints them.

In special cases, where it shall be deemed necessary, the adjudicating corporation shall have the right of appointing a specialist to take part in the deliberations and decisions of a Nobel-Committee, in the capacity of a member of the same.

§ 7.

It is essential that every candidate for a prize under the terms of the Will be proposed as such in writing by some duly qualified person. A direct application for a prize will not be taken into consideration.

The qualification entitling a person to propose another for the receipt of a prize consists in being a representative, whether Swedish or otherwise, of the domain of Science, Literature &c in question, in accordance with the detailed stipulations obtainable from the corporations charged with adjudicating the prizes.

At each annual adjudication those proposals shall be considered that have been handed in during the twelve months preceding the 1st day of February.

§ 8.

The grounds upon which the proposal of any candidate's name is made must be stated in writing and handed in along with such papers and other documents as may be therein referred to.

Should the proposal be written in a language other than those of the Scandinavian group, or than English, French, German or Latin, or should the adjudicators, in order to arrive at a decision upon the merits of a work proposed, be under the necessity of obtaining information as to the contents chiefly from a work written in a language, for the understanding of which there is no expedient save such as involves a great expenditure of trouble or money, it shall not be obligatory for the adjudicators to pay further consideration to the proposal.

§ 9.

On Founder's Day, the 10th of December, the anniversary of the death of the testator, the adjudicators shall make known the results of their award and shall hand over to the winners of prizes a cheque for the amount of the same, together with a diploma and a medal in gold bearing the testator's effigy and a suitable legend.

It shall be incumbent on a prize-winner, wherever feasible, to give a lecture on the subject treated of in the work to which the prize has been awarded; such lecture to take place within six months of the Founder's Day at which the prize was won, and to be given at Stockholm or, in the case of the Peace prize, at Christiania.

§ 10.

Against the decision of the adjudicators in making their award no protest can be lodged. If differences of opinion have occurred they shall not appear in the minutes of the proceedings, nor be in any other way made public.

§ 11.

As an assistance in the investigations necessary for making their award, and for the promotion in other ways of the aims of the Foundation, the adjudicators shall possess powers to establish scientific institutions and other organizations.

The institutions &c so established, and belonging to the Foundation, shall be known under the name of Nobel-Institutes.

§ 12.

Each of the Nobel-Institutes shall be under the control of that adjudicating corporation that has established it.

As regards its external management and its finances a Nobel-Institute shall have an independent status. Its property is not, however, on that account available for defraying the expenses of any establishments belonging to an adjudicating or any other corporation. Nor is it permissible for any scholar who is in receipt of a fixed salary as an official of a Swedish Nobel-Institute to occupy a similar position at any other institution at the same time, unless the King be pleased to permit it in a special case.

So far as the adjudicators of prizes deem it to be feasible, the Nobel-Institutes shall be established on one common site and shall be organised uniformly.

The adjudicating corporations are at liberty to appoint foreigners, either men or women, to posts at the Nobel-Institutes.

§ 13.

From that portion of the income derived from the main fund that it falls to the lot of each of the five Sections annually to distribute, one fourth of the amount shall be deducted before the distribution is made. The immediate expenses connected with the award having been discharged, the remainder of the amount deducted as above directed shall be employed to meet the expenses of the Section in maintaining its Nobel-Institute. The money which is not absorbed in thus defraying the current expenditure for the year, shall form a reserve fund for the future needs of the Institute.

THE ADMINISTRATION OF THE FOUNDATION.

§ 14.

The Nobel-Foundation shall be represented by a Board of Control, located in Stockholm. The Board shall consist of five members, one of whom, the President, shall be appointed by the King, and the others by the delegates of the adjudicating corporations. The Board shall elect from their own members a Managing Director.

For the member of the Board whom the King appoints one substitute shall be chosen, and for the other members two substitutes.

Those members of the Board who are elected by the delegates of the adjudicators, and also their substitutes, shall be appointed to hold office for two years, commencing from the 1st day of May.

§ 15.

The Board shall administer the funds of the Foundation as well as the other property, real and otherwise, belonging to it, in so far as such is common to all the sections.

It shall be a function of the Board to hand over to the winners of prizes in accordance with the rules of the Foundation, the prizes so won, and besides, to attend to the payment of all duly authorised expenses connected with the prize-distribution, the Nobel-Institutes and similar objects. It shall further be incumbent on the Board to be of assistance, in matters that are not of a scientific character, to all those who have to do with the Foundation, where help be required.

The Board shall be empowered to engage the services of a lawyer to summon or prosecute a person or to defend a case on its behalf if need arise, and, in general, to act as the legal representative of the Foundation. The Board shall be entitled to engage the assistants who may be necessary for the proper discharge of its duties, and also to fix the terms, both as regards salary and pension, on which such assistants shall be remunerated.

§ 16.

The adjudicating corporations shall appoint fifteen delegates, for two civil years at a time; of these delegates the Academy of Science shall choose six and each of the other bodies three. To provide against inconvenience from the disability of a delegate to serve at any time, the Academy of Science shall appoint four substitutes, and each of the other bodies two.

The delegates shall elect one of their number to act as chairman. That election shall be held at a meeting to which the oldest of the delegates chosen by the Academy of Science shall summon his fellow-delegates.

A minimum of nine delegates shall constitute a quorum. If any of the adjudicating corporations neglect to choose delegates, that shall not prevent the other delegates from arriving at a decision on the business before them.

Should a delegate reside at any place other than that where the meeting of delegates takes place, he shall be entitled to receive reasonable compensation for the expense to which he shall have been put in attending the meeting, such compensation to be paid from the general funds of the Foundation.

§ 17.

The administration and accounts of the Board shall be controlled once every civil year by five auditors, of whom each of the adjudicating corporations shall elect one and the King appoint the fifth; this last shall act as chairman at their sittings.

Before the expiration of February every year a report concerning the administration of the Board shall be handed in to the chairman of the Auditing Committee, which in its turn shall bring in its report before the first day of April to the delegates of the adjudicating corporations.

In the Auditors' Report, which must be published in the public newspapers, there shall appear a summary of the objects to which the proceeds of the several funds have been applied.

If any of the adjudicating corporations neglects to elect an auditor, or if any auditor fails to appear after having been summoned to a sitting of the Auditing Committee, the other auditors shall not be thereby prevented from pursuing their task of auditing.

§ 18.

The auditors shall at all times have access to all the books, accounts and other documents of the Foundation; nor shall any information they may demand concerning the management be withheld by the Board. All the deeds and securities belonging to the Foundation shall be examined and verified at least once a year by the auditors.

The Minister of Public Education and Worship, either in person or by appointed deputy, shall also have the right of access to all the documents belonging to the Foundation.

§ 19.

On the basis of the Auditors' Report the delegates of the adjudicators shall determine whether the Board shall be held absolved from their responsibility or not, and shall take those measures against the Board or any member of it for which call may arise. If no case be brought up within a year and a day of the date when the report of the Board was handed in to the auditors, the exoneration of the Board shall be held to have been granted.

§ 20.

The King shall determine the salary of the managing director, and also the amount of remuneration that shall be given to the other members of the Board and to the auditors.

Further instructions as to the management of the Foundation not contained in this Code shall be issued by the King in special by-laws.

§ 21.

One tenth part of the annual income derived from the main fund shall be added to the capital. To the same fund shall be also added the interest accruing from the sums set aside for prizes, while they remain undistributed or have not been carried over to the main or other (special) fund, as directed in § 5.

ALTERATIONS IN THE CODE.

§ 22.

A proposition to modify these statutes may be made by any of the adjudicating corporations, by their delegates, or by the Board. Upon any such

proposition being brought forward by the adjudicators or by the Board, the delegates shall be required to express an opinion relative to it.

The adjudicators and the Board shall have to come to a decision on any
proposal made, the Academy of Science having two votes and the other corporations
one each. If there are not at least four votes in favour of a proposition, or if
that corporation whose rights and authority the change proposed affects has not
given its assent, the proposition shall be regarded as rejected. In the contrary
case the proposition shall be submitted by the Board to the King for his
consideration.

The omission on the part of any of those who are notified in due course
of a proposed change, to send in any communication within four months of the receipt
of the said notification, shall not prevent a decision being arrived at.

TEMPORARY REGULATIONS.

1. Directly the Code of Statutes of the Foundation shall have been ratified
by the King, the adjudicators shall appoint the prescribed number of delegates
to act until the close of the year 1901; they shall be summoned to meet together
in Stockholm at the earliest date possible, for the purpose of electing the members
of the Board of Control of the Foundation.

In determining the period of service of those members of the Board who
are first appointed, the following points are to be observed: firstly, that to the
time of service laid down by the statutes, which commences on 1st May 1901, the
time between the date of the election and the day named must be added, and
secondly, that two members of the Board shall be chosen by lot to go off again
one year afterwards (on May 1).

2. The Board of Control of the Foundation shall assume the management
of the property of the Foundation at the commencement of the year 1901;
subject to the proviso, however, that the testator's executors shall be at liberty
to continue, during the progress of the year, to take those measures which may
still be necessary for the completion of the winding up of the estate, so far as
they find needful.

3. The first distribution of prizes shall take place, if feasible, in 1901, and
that in all five sections.

4. From the property possessed by the Foundation there shall be deducted:
   a. A sum of 300,000 kronor (about £ 16,556) for each of the five sections,
   1,500,000 kronor in all, to be used, along with the interest accruing
   therefrom after the first of January 1900, as need arises, for defraying
   the running expenses of organising the Nobel-Institutes, and
   b. The sum which the Board, after consultation with the delegates, may
   deem necessary for procuring a building of its own, to embrace offices
   for the transaction of business and a large hall for Founder's-day celebrations.

The adjudicators shall be empowered to set aside the 300,000 kronor and
interest thereon, mentioned above, or any portion of the same, on behalf of the
special funds of the different sections.

To all which Each and Every One, whom it may concern, hath to pay dutiful and obedient heed.   To the further certainty whereof WE have hereby attached OUR own signature and royal seal.

At the Palace in Stockholm, on this the 29th day of June 1900.

OSCAR.

(L. S.)

SPECIAL REGULATIONS,

CONCERNING

THE DISTRIBUTION ETC. OF PRIZES
FROM THE NOBEL FOUNDATION

BY

THE ROYAL ACADEMY OF SCIENCE IN STOCKHOLM.

GIVEN BY

HIS GRACIOUS MAJESTY, OSCAR II, KING OF SWEDEN AND NORWAY,
AT THE PALACE IN STOCKHOLM, ON THE 29TH DAY OF JUNE 1900.

PRIZE DISTRIBUTION.

§ 1.

The right to hand in the name of a candidate for a prize, as directed in § 7 of the Code of Statutes of the Nobel-Foundation, shall belong to:

1. Home and foreign members of the Royal Academy of Science in Stockholm.

2. Members of the Nobel-Committees of the Physical and Chemical Sections as defined in the Code.

3. Scientists who have received a Nobel-prize from the Academy of Science.

4. Professors, whether in ordinary or associate, of the Physical and Chemical Sciences at the Universities of Upsala, Lund, Christiania, Copenhagen and Helsingfors, at the Caroline Medico-Chirurgical Institute and the Royal Technical College in Stockholm, and also those teachers of the same subjects who are on the permanent staff of the Stockholm University College.

5. Holders of similar chairs at other universities or university colleges, to the number of at least six, to be selected by the Academy of Science in the way most appropriate for the just representation of the various countries and their respective seats of learning.

6. Other Scientists whom the Academy of Science may see fit to select.

A determination as to the choice to be made of teachers and scientists, in accordance with sections 5 and 6 above, shall be arrived at before the close of each September.

§ 2.

For each of the Physical and Chemical Sections the Nobel-Committee, as prescribed in § 6 of the Code, shall consist of five members, four of them being chosen by the Academy and the fifth being the president of the corresponding section of the Nobel-Institute, as mentioned in § 14 below.

The election shall be for a space of four civil years. A member going off by rotation shall be eligible for reelection.

If a member retires or dies before his period expires, another person shall be elected to serve for the remainder of the period.

§ 3.

Previous to the election of a member of the Nobel-Committee, a list of proposed names shall be drawn up by the 4th Class in the Academy if the election be to the Committee in Physics, and by the 5th Class if the election be to the Committee in Chemistry. These lists shall be handed in to the Academy not later than the close of November.

If either of the above Classes of the Academy so desires, they shall be empowered to associate any competent member of another Class with themselves in the task of drawing up the lists aforesaid.

§ 4.

The Academy shall select one of the members chosen to sit on a Nobel-Committee to be the chairman of the same, for the space of one year at a time. In case of absence on the part of the chairman, his place shall be taken for the sitting by the oldest among the members present.

When the two committees meet in joint conference the chair shall be taken by the older of the two chairmen.

§ 5.

No decision shall be arrived at by a Nobel-Committee, unless there be present a minimum of three out of the five members having seats on it, as directed in § 2 above.

Voting shall not be by ballot, but open. If the votes be equally divided, the chairman shall have a casting vote.

§ 6.

During the course of the month of September in each year the Nobel-Committees shall issue a circular to all those who are qualified, according to § 1 above, summoning them to make nominations of candidates for prizes before the first day of February in the following year; such nominations to be supported by evidence, documentary and otherwise.

§ 7.

Before the close of September every year the Nobel-Committee shall present to the Academy their opinion and proposals regarding the distribution of prizes.

That Class in the Academy which is therein concerned shall then express its views with regard to the proposals, before the expiration of the month of October at the latest. Should the Class in question deem it necessary to call in the services of some qualified member of any other Class, to aid in drawing up their report, they shall have authority to do so.

The final decision, devolving upon the Academy, shall be arrived at within the lapse of the first half of November next ensuing.

§ 8.

The proceedings, verdicts and proposals of the Nobel-Committees with reference to the prize-distribution shall not be published or in any other way be made known.

§ 9.

The amount of the remuneration that in conformity with § 6 in the Code is to be allotted to a member of a Nobel-Committee, shall be determined by the Academy, after it has heard the joint views of Classes 5 and 6.

The amount of remuneration to be accorded to any person who shall have been called in as an expert member of a Nobel-Committee, in pursuance of the stipulations of § 6 in the Code, shall be determined by the Academy, after it has heard the opinion of the Class which shall have called in such member.

§ 10.

To every member of the Academy who shall attend a meeting at which, in pursuance of § 7 (item 2 or 3), a Class in the Academy shall agree upon a final verdict or at which the Academy shall come to a decision in regard to the prize-award, a Nobel medal in gold shall be presented for each occasion.

§ 11.

All questions connected with the Nobel-Foundation shall be dealt with at special sittings of the Academy. The minutes made at those sittings shall not be preserved with those of the other sittings of the Academy. All expenses entailed by these special sittings shall be defrayed from the funds of the Nobel-Foundation.

THE NOBEL INSTITUTE.

§ 12.

The Nobel-Institute, which § 11 of the Code authorises the Academy of Science to establish, is to be so established primarily for the purpose of carrying out, where the respective Nobel-Committees shall deem requisite, scientific investigation as to the value of those discoveries in the domains of Physics and Chemistry, which shall have been proposed as meriting the award of a Nobel-prize to their authors.

The Institute shall, moreover, as far as its means allow, promote such researches in the domains of the sciences named, as promise to result in salient advantage.

§ 13.

The Nobel-Institute shall consist of two sections, one for Physical Research and one for Chemical Research.

The buildings required for these two sections shall be erected on contiguous sites, and rooms for the sittings of the Nobel-Committees as well as record-rooms, libraries &c shall be constructed for the two in common.

§ 14.

The Nobel-Institute shall be under the superintendence of an Inspector, appointed by the Crown.

As president of each of the two sections of the Nobel-Institute, the Academy of Science shall select, on the basis of recommendations from the Class in the Academy concerned, a scientist, either of Swedish or foreign extraction, who is possessed of an established reputation as an investigator and of a wide experience in, and grasp of, the branch of science which it is the function of the section to promote.

The presidents shall have the title of Professor.

The terms of appointment for the presidents shall be drawn up by the Academy on the basis of suggestions from the Class in the Academy concerned.

§ 15.

The president of a section shall devote the whole of his working-time to the concerns of that section. He shall exercise supervision over the officials and attendants in the service of the section, have charge of the buildings and collections belonging to it, and be held responsible in the last resort for the finances.

The president shall see to the carrying out of the work of investigation mentioned in § 12. In cases where such work falls within that department of research which the president has made his own, he shall be required to execute it himself.

The other regulations to which the president shall be subject shall be imparted to him in a special code of instructions drawn up by the Academy.

§ 16.

Whenever need shall arise for the calling in of a specialist to assist in the work of investigation, that Nobel-Committee which has the matter in hand, shall make application to the Academy for the purpose. The fee for such work shall be fixed in each case by the Academy on the basis of the Committee's own proposal, observance nevertheless being paid to the following paragraph — § 17.

§ 17.

In cases where the Academy, by the terms of the Code, does not hold the sole right to determine the amount of the remuneration to be accorded to any member of the Academy, the decision authorising such payments to be made shall be submitted to the Crown for consideration and sanction.

§ 18.

A joint secretary for the two sections of the Nobel-Institute shall be appointed by the Academy, the conditions of appointment to rest with that body. Names for the post shall be proposed by the two Nobel-Committees jointly. The secretary shall be required, in addition to his other duties, to keep the minutes at the sittings of the Nobel-Committees. A librarian shall also be appointed in the same manner. The position of librarian may be combined with that of secretary or assistant to the Institute.

Assistants, makers of instruments, porters and other officials required for the work of the Institute, shall be engaged and dismissed by that Nobel-Committee which employs them.

§ 19.

Permission for other persons than those who are on the scientific staff of the Institute to carry on research in its laboratories &c, may be granted by the Nobel-Committee interested, yet only provided the researches are directed towards determining the scientific conditions upon which some discovery or some invention may be evolved.

SPECIAL FUNDS.

§ 20.

As soon as any Special Funds shall have been formed, in accordance with § 5 in the Code, the Academy shall be entitled to distribute, out of the annual yield thereof, support for the furtherance, in directions the testator had ultimately in view in making his bequest, of any work in the domains of Physical and Chemical Science that may be judged to be of significance either in a scientific or a practical regard.

Assistance of that kind shall by preference be accorded to such persons as shall have already attained, by their labours in the sciences named, to results that promise in their further development to prove worthy of the support of the Nobel-Foundation.

Proposals for the awarding of assistance of the nature above indicated shall be made by the respective Nobel-Committees and submitted to the Academy; it shall then rest with that body to consult the opinion of the Class concerned and thereafter to determine on the case.

The income derived from the special funds may also be applied to the needs of the Nobel-Institute.

ALTERATION OF THE PRESENT STATUTES.

§ 21.

A proposition to alter the present statutes may be raised by any member of the Academy or of the Nobel-Committees. Before the Academy proceeds to deal with any proposition to that end, it shall first obtain an expression of opinion with regard to it from the two Nobel-Committees jointly,· and subsequently from Classes 4 and 5 in the Academy jointly. Any proposed alteration that has been adopted by the Academy shall be submitted to the Crown for consideration and sanction.

TEMPORARY REGULATIONS.

On the occasion of the first election of members on the Nobel-Committees the Academy shall also appoint a pro tem. secretary for these Committees.

Until such time as presidents shall have been chosen or definitely appointed, there shall be a fifth member of each of the Nobel-Committees, chosen by the Academy. Those members shall retire on the appointment of presidents.

In determining the period of service of the other four members of each Committee to be first appointed, the following points are to be noted: that to the period stipulated for them to act must be added the time that elapses between the day of their election and the 1st of January 1901; and further, that at the time of election lots shall be drawn to determine which of the members shall go off by rotation, as stipulated, at the close of the years 1901, 1902 and 1903.

The presidents of the sections of the Institute shall be appointed pro tem., directly after the Academy has decided that measures shall be taken for the establishment of the Institute.

The definite appointment to the permanent posts of both president and secretary shall not take place until the Institute shall have been equipped and be in working order.

Until the time when the Nobel-Institute shall be complete and have obtained its due organization, the Nobel-Committees shall resort to the opinions of experts in the several departments for such technical information as they may find themselves in need of for the purposes of the adjudicating of prizes, and they are empowered to have the experimental investigation and testing carried out at any institution, either home or foreign, that they may deem suitable. The fees to be paid in such cases shall be individually fixed by the Academy on the basis of a suggestion to be made by the Nobel-Committee concerned, with due observance, however, of the stipulations contained in § 17.

To all which Each and Every One, whom it may concern, hath to pay dutiful and obedient heed. To the further certainty whereof WE have hereby attached OUR own signature and royal seal.

At the Palace in Stockholm, on this the 29th day of June 1900.

OSCAR.

(L. S.)

Members of the Nobel committees for physics and chemistry elected 1900–1915

NOBEL COMMITTEE FOR PHYSICS

K. Ångström 1900–1910	Professor of Physics at Uppsala University
S. Arrhenius 1900–1927	Professor of Physics at the Stockholm Högskola; subsequently Director, Nobel Institute for Physical Chemistry
V. Carlheim-Gyllensköld 1910–1934	Professor of Physics at the Stockholm Högskola
G. Granqvist 1904–1922	Professor of Physics at Uppsala University
A. Gullstrand 1911–1929	Professor of Physiology and Physical Optics at Uppsala University
B. Hasselberg 1900–1922	Physicist at the Royal Academy of Sciences
H. H. Hildebrandsson 1900–1910	Professor of Meteorology at Uppsala University
R. Thalén 1900–1903	Professor of Physics at Uppsala University

NOBEL COMMITTEE FOR CHEMISTRY

P. T. Cleve 1900–1905	Professor of Chemistry at Uppsala University
Å. G. Ekstrand 1913–1924	Engineer, civil servant

O. Hammarsten
1905–1926

Professor of Physiological Chemistry at Uppsala University

P. Klason
1900–1925

Professor of Chemistry and Chemical Technology at the Royal Institute of Technology

O. Pettersson
1900–1912

Professor of Chemistry at the Stockholm Högskola

H. G. Söderbaum
1900–1933

Professor of Agricultural Chemistry at the Experimental Agricultural Station (Academy of Agriculture)

O. Widman
1900–1928

Professor of Chemistry at Uppsala University

Nobel prizewinners in physics and chemistry, 1901–1915

PHYSICS

1901 *Wilhelm Conrad Röntgen:* In recognition of the extraordinary services he has rendered by the discovery of the remarkable rays subsequently named after him.

1902 *Hendrik Antoon Lorentz and Pieter Zeeman:* In recognition of the extraordinary service they have rendered by their researches into the influence of magnetism upon radiation phenomena.

1903 *Henri Becquerel:* In recognition of the extraordinary services he has rendered by his discovery of spontaneous radioactivity.

Pierre Curie and Marie Curie: In recognition of the extraordinary services they have rendered by their joint researches on the radiation phenomena discovered by Professor Henri Becquerel.

1904 *Lord Rayleigh (J. W. Strutt):* For his investigations of the densities of the most important gases and for his discovery of argon in connection with these studies.

1905 *Philipp Lenard:* For his work on cathode rays.

1906 *Joseph John Thomson:* In recognition of the great merits of his theoretical and experimental investigations on the conduction of electricity by gases.

1907 *Albert Abraham Michelson:* For his optical precision instruments and the spectroscopic and metrological investigations carried out with their aid.

1908 *Gabriel Lippmann:* For his method of reproducing colors photographically based on the phenomenon of interference.

1909 *Gugliemo Marconi and Ferdinand Braun:* In recognition of their contributions to the development of wireless telegraphy.

1910 *Johannes Diedrik van der Waals:* For his work on the equation of state for gases and liquids.

1911 *Wilhelm Wien:* For his discoveries regarding the laws governing the radiation of heat.

1912 *Nils Gustaf Dalén:* For his invention of automatic regulators for use in conjunction with gas accumulators for illuminating lighthouses and bouys.

1913 *Heike Kamerlingh Onnes:* For his investigations on the properties of matter at low temperatures which led, inter alia, to the production of liquid helium.

1914 *Max von Laue:* For his discovery of the diffraction of X-rays by crystals.

1915 *William Henry Bragg and William Lawrence Bragg:* For their services in the analysis of crystal structure by means of X-rays.

CHEMISTRY

1901 *Jacobus Henricus van't Hoff:* In recognition of the extraordinary services he has rendered by the discovery of the laws of chemical dynamics and osmotic pressure in solutions.

1902 *Emil Fischer:* In recognition of the extraordinary services he has rendered by his work on sugar and purine syntheses.

1903 *Svante Arrhenius:* In recognition of the extraordinary services he has rendered to the advancement of chemistry by his electrolytic theory of dissociation.

1904 *William Ramsay:* In recognition of his services in the discovery of the inert gaseous elements in air and his determination of their place in the periodic system.

1905 *Adolf von Baeyer:* In recognition of his services in the advancement of organic chemistry and the chemical industry, through his work on organic dyes and hydroaromatic compounds.

1906 *Henri Moissan:* In recognition of the great services rendered by him in his investigation and isolation of the element fluorine, and for the adoption in the service of science of the electric furnace named after him.

1907 *Eduard Buchner:* For his biochemical researches and his discovery of cell-free fermentation.

1908 *Ernest Rutherford:* For his investigations into the disintegration of the elements and the chemistry of radioactive substances.

1909 *Wilhelm Ostwald:* In recognition of his work on catalysis and for his investigations into the fundamental principles governing chemical equilibria and rates of reaction.

1910 *Otto Wallach:* In recognition of his services to organic chemistry and the chemical industry by his pioneer work in the field of alicyclic compounds.

1911 *Marie Curie:* In recognition of her services to the advancement of chemistry by the discovery of the elements radium and polonium, by the isolation of radium, and the study of the nature and compounds of this remarkable element.

1912 *Victor Grignard:* For the discovery of the so-called Grignard reagent, which in recent years has greatly advanced the progress of organic chemistry.

Paul Sabatier: For his method of hydrogenating organic compounds in the presence of finely disintegrated metals whereby the progress of organic chemistry has been greatly advanced in recent years.

1913 *Alfred Werner:* In recognition of his work on the linkage of atoms in molecules by which he has thrown new light on earlier investigations and opened up new fields of research especially in inorganic chemistry.

1914 *Theodor William Richards:* In recognition of his accurate determinations of the atomic weight of a large number of chemical elements.

1915 *Richard Willstätter:* For his researches on plant pigments, especially chlorophyll.

A note on sources and notes

THE NOBEL ARCHIVES OF THE ROYAL ACADEMY OF SCIENCES

The primary source for this study is the archives of the Royal Academy of Sciences and its Nobel Committees for Physics and Chemistry. In accordance with the new rule that was instituted in 1974 (para. 10 of the statutes), any of the four institutions awarding the Nobel prizes can permit access to its records for purposes of historical research. Such permission may be given for prize decisions dating back fifty years or more.

The archives that concern the prizes in physics and chemistry are located at the Academy of Sciences and at the joint office of its two committees in the Nobel Foundation building in Stockholm. Thanks to a generous grant from Stiftung Volkswagenwerk, it has been possible to microfilm the materials in the archives for the period 1900–1929, the latter year being the last for which permission to consult the archives was granted at the time of the microfilm project. A copy of the microfilm has been deposited at the National Record Office (Riksarkivet) for use by scholars who have been granted permission by the Academy to consult materials in the archives.

Presented in the order in which they entered the process of decision making about the prizes in physics and chemistry, the most important materials making up the archives are the following (for a detailed description of these materials see E. Crawford and R. MacLeod, *The Nobel prizes in physics, 1901–1916: A report on archival materials* [Paris and Brighton, 1976]):

> 1. *Letters of nomination.* These are the letters addressed to the two committees in response to invitations to nominate candidates for the prizes.

242

The originals of the letters are preserved at the office of the committees, but copies were also made each year and sent to the Academy with the committees' recommendation for the year's prizewinner.

2. *Reports on the candidates.* There are two types of reports: (a) the *special reports*, which contain detailed assessments of the work of leading contenders for the prizes of each year; and (b) the *general reports*, where all the candidates of the year are reviewed and short statements made as to the merit of the work of each of them. The drafts of the special reports (found in the committees' archives) formed the basis of the recommendation for the year's prizewinner that the committees made to the Academy. The final typescripts of the general and special reports handed in to the Academy constitute annexes to the minutes of Academy meetings concerning the Nobel prizes.

3. *Minutes of the Nobel Committees for Physics and Chemistry.* Recorded here are only the votes taken and the decisions reached with respect to the committees' recommendations of prizewinners, but not the discussions leading up to the decisions. The minutes are found in the committees' archives.

4. *Minutes of meetings in the Academy concerning the Nobel prizes.* These are, on the one hand, the minutes of the physics and chemistry sections and, on the other, those of the Academy meeting in plenary session. In the minutes of the sections, the votes taken on the sections' recommendations to the Academy concerning the prizewinners are recorded but not the discussions. In the minutes of the Academy meeting only the final decision is recorded.

In the Notes, the documents listed under items 1–4 above are referred to as follows:

1. *Förslag, fysik/kemi = Förslag till utdelning av Nobelpriset i fysik/kemi* (letters of nomination concerning the awarding of the Nobel prizes in physics or chemistry).

2. (a) *KU, fysik/kemi = Kommittéutlåtande, Nobelkommittén för Fysik/Kemi* (committee report, Nobel Committee for Physics or Chemistry). Refers to general and special reports as handed into the Academy. (b) *Utkast, KU, fysik/kemi = Utkast, Kommittéutlåtande, Nobelkommittén för Fysik/ Kemi* (draft, committee report, Nobel Committee for Physics or Chemistry). Refers to drafts of special reports prepared for the committees. The dates given in the notes are those of the meetings when the reports were handed in to the committee.

3. (a) *Protokoll, NK, fysik/kemi = Protokoll vid Kungl. Vetenskapsakademiens Nobelkommittés för Fysik/Kemi sammanträde* (minutes of the meetings of the Nobel Committees for Physics or Chemistry). (b) *Protokoll, gem. sam. = Protokoll vid Kungl. Vetenskapsakademiens fysiska och kemiska Nobelkommittéers gemensamma sammanträden* (minutes of the joint meetings of the Nobel Committees for Physics and Chemistry).

4. (a) *Nobelprotokoll, KVA* and (b) *Nobelprotokoll, KVA, 3dje/4de klassen = Protokoll vid Kungl. Vetenskapsakadamiens sammankomster för behandling av ärenden rörande Nobel-prisen* (minutes of meetings of the Royal Swedish Academy of Sciences for discussion of matters concerning the Nobel prizes). (a) Refers to plenary meetings of the Academy; (b) refers to meetings of its sections for physics and chemistry.

The minutes of the regular meetings of the Academy and its sections are referred to as *Protokoll, KVA* (minutes of meetings of the Royal Swedish Academy of Sciences) and *Protokoll, KVA, 3dje/4de klassen* (minutes of meetings of the Academy sections for physics and chemistry).

Other abbreviations used in the Notes are the following:

ANS	Alfred Nobels Sterbhus Arkiv, Ragnar Sohlmans handlingar (Archive of the Estate of Alfred Nobel, documents belonging to Ragnar Sohlman)
ANT	Alfred Nobels Testamentsexekutorers samt Nobelkommitténs Arkiv (Archive of the Executors of the Estate of Alfred Nobel and of the Large Nobel Committee)
CHP/AIP	Center for the History of Physics, American Institute of Physics, New York
DSB	*Dictionary of Scientific Biography*
HSPS	*Historical Studies in the Physical Sciences*
K. or Kungl.	Kungliga (Royal)
KB	Kungl. Biblioteket (Royal Library), Stockholm
KVA	Kungl. Vetenskapsakademien (Royal Academy of Sciences), Stockholm
KVÅb	*Kungl. Vetenskapsakademiens årsbok (Yearbook of the Royal Academy of Sciences)*
LC	Library of Congress, Washington, D.C.

M.-L. Inst.	Makarna Mittag-Lefflers Matematiska Stiftelse (Mittag-Leffler Institute), Djursholm
OUB	Universitetsbiblioteket (University Library), Oslo
RA	Riksarkivet (National Record Office), Stockholm
SA	Statsarkivet (City Archives), Stockholm
SUB-KVAB	Stockholms Universitetsbibliotek med Kungl. Vetenskapsakademiens Bibliotek (Stockholm University Library with the Library of the Royal Academy of Sciences), Stockholm
UUB	Uppsala Universitetsbibliotek (Uppsala University Library), Uppsala.

OTHER SOURCES

Among other sources for the history of the Nobel prizes that have not been used previously is the important collection of newspaper clippings held by the Nobel Foundation. This collection was begun in 1896 when the executors to the estate retained a Paris clipping agency to monitor the world press and to forward all items making any reference to Nobel or the Nobel prizes. The clippings were pasted into bound volumes of one hundred pages each. For the period 1896–1915, the collection consists of sixty-four volumes of clippings from the Swedish and foreign press. Most of the journal articles cited in the Notes emanate from this collection. The collection has been microfilmed for the years 1896–1930, and copies of the films have been deposited at the Nobel Foundation. The microfilming was carried out with the generous assistance of the Centre National de la Recherche Scientifique and Stiftung Volkswagenwerk.

Two archives were particularly important for drawing up the account of the negotiations over the will of Alfred Nobel and the statutes of the Nobel Foundation: (1) the Archive of the Estate of

Alfred Nobel (documents belonging to Ragnar Sohlman), found at the National Record Office; and (2) the Archive of the Executors of the Estate of Alfred Nobel and of the Large Nobel Committee (papers belonging to Carl Lindhagen), found at the Stockholm City Archives. The materials of most interest for the present study were those of the latter archive.

In addition, a large number of other archival sources – in particular, the personal papers of Swedish scientists – are referred to in the Notes.

NOTES

Published works cited in the Notes are referred to by author's name, title, city and year of publication when they appear for the first time; after that only the name of the author and a short title are used. Full references to the works cited are found in the Bibliography.

Notes

Introduction

1. On the importance of recognition and symbolic rewards in science and the forms these take, see R. K. Merton, *The sociology of science: Theoretical and empirical investigations* (Chicago, 1973), pp. 297–302; J. R. Cole and S. Cole, *Social stratification in science* (Chicago, 1973), pp. 46–60 and Appendix B.

2. In Nobel's earlier will, dated 1893, the Royal Academy of Sciences was the sole institution designated to award an unspecified number of prizes "for the most important and pioneering works within the wide domain of knowledge and progress." It was to have received 64 percent of Nobel's fortune, the interest on which was to be used for the prizes.

3. This genre has dominated the vast literature on the Nobel prizes (a far from exhaustive bibliography compiled early in this project ran to three hundred references). The two most important exceptions to this approach are: (1) the official history of the institution (H. Schück et al., *Nobel: The man and his prizes* [Amsterdam, 1962]), which is of interest mainly because the lengthy articles on the prizes in science and medicine place these in the context of the development of the disciplines concerned (but generally overstate the importance of the prizes for such developments); and (2) the exhaustive sociological study of American prizewinners conducted by Zuckerman (H. Zuckerman, *Scientific elite: Nobel laureates in the United States* [New York, 1977]). The extent to which the general genre has taken hold is illustrated by the reviewer who felt that inside Zuckerman's book there was another one struggling to get out – "a free-ranging discourse on the Nobel prizes in general" (D. Fleming, review of Zuckerman's *Scientific elite, Times Literary Supplement* [Februry 3, 1978]).

4. Preliminary results of these studies, some of which are still under way, were presented at Nobel Symposium No. 52, "Science, technology, and society in the time of Alfred Nobel" (C. G. Bernhard, E. Crawford, and P. Sörbom, *Science, technology, and society in the time of Alfred Nobel* [Oxford, 1981], pp. 307–403).

5. Zuckerman, *Scientific elite*, pp. 11–13.

6. See, for example, M. Berman, *Social change and scientific organization: The Royal Institution, 1799–1844* (London, 1978); R. Hahn, *The anatomy of a scientific institution: The Paris Academy of Sciences, 1666–1803* (Berkeley, Calif., 1971); J. Morrell and A. Thackray, *Gentlemen of science: Early years of the British Association for the Advancement of Science* (Oxford, 1981).

7. P. Forman, J. Heilbron, and S. Weart, "Physics *circa* 1900: Personnel, funding, and productivity of the academic establishments," *HSPS*, 5 (1975), p. 12.

8. At the turn of the century, the U.S. dollar was worth about 4 Swedish crowns and the British pound about 18 crowns. Approximately the same ratios applied to the German mark. The corresponding figures for the French franc were 5 and 25, respectively. (Forman et al., "Physics *circa* 1900," p. 5.) In terms of its

purchasing power at the time, the Nobel prize of 1901 (150,000 crowns) was approximately double that of 1983 (1.5 million crowns).

CHAPTER 1 *Precursors to the Nobel prizes in the sciences*

1. *The Record of the Royal Society of London* (London, 1897), pp. 124–5.
2. E. Maindron, *Les fondations de prix à l'Académie des Sciences: Les lauréats de l'Académie, 1714–1880* (Paris 1881), pp. 13–17.
3. A. Harnack, *Geschichte der Königlich Preussischen Akademie der Wissenschaften zu Berlin* (Berlin, 1900), Vol. I, pp. 396–7.
4. Ibid., p. 287.
5. Maindron, *Les fondations de prix*, pp. 26–8, 36; S. Lindroth, *Kungl. Svenska Vetenskapsakademiens historia, 1739–1818* (Stockholm, 1967), Vol. I, pp. 140 ff.
6. This designation indicated that the subject for competition was set in alternate years by the section of mathematical sciences and the section of physical sciences.
7. R. Fox, "The rise and fall of Laplacian physics," *HSPS*, 4 (1974), p. 103.
8. Maindron, *Les fondations de prix*, p. 57; J. T. Merz, *A history of European scientific thought in the nineteenth century* (Gloucester, Mass., 1976), Vol. II, pp. 25 ff.
9. Harnack, *Geschichte*, Vol. I, pp. 792, 898.
10. C. de Comberousse, *J.-B. Dumas (1800–1884)* (Paris, 1884), pp. 24–5.
11. Maindron, *Les fondations de prix*, p. 39.
12. M. Crosland, "From prizes to grants in the support of scientific research in France in the nineteenth century: The Montyon legacy," *Minerva*, 17 (1979), p. 361.
13. Crosland, "From prizes to grants," pp. 367–8.
14. The years given in parentheses are those when the donations were made public, usually on the occasion of the death of the donor. The Alberto Levi prize was awarded E. von Behring and E. Roux in 1895 for the application of serum therapy to diphtheria. P. Gauja, *Les fondations de l'Académie des Sciences (1881–1915)* (Hendaye, France, 1917), p. 548; *Nobel Foundation directory, 1979–1980* (Stockholm, 1979), p. 15.
15. To give an example from another area, the *prix Pierre Guzman* (1899) provided for an award of 100,000 francs to the person who found a way to communicate with a celestial body. The terms of the bequest stated that as long as a solution to this problem had not been found, the interest accumulated on this sum would be used for a prize in general astronomy. Gauja, *Les fondations de l'Académie*, p. 404.
16. R. MacLeod, "Of medals and men: A reward system in Victorian science, 1826–1914," *Notes and Records of the Royal Society of London*, 26 (1971), pp. 81–105.
17. Gauja, *Les fondations de l'Académie*, pp. 485–506.
18. G. Darboux, *Eloges académiques et discours* (Paris, 1912), p. 302.
19. Darboux, *Eloges*, p. 300.
20. Forman et al., "Physics *circa* 1900," pp. 49 and 76.
21. *Sitzungsberichte der Königlich Preussischen Akademie der Wissenschaften, 1914*, pp. 80–1.
22. *The Record of the Royal Society*, pp. 120–30.
23. R. MacLeod, "The support of Victorian sciences: The endowment of research movement in Great Britain, 1868–1900," *Minerva*, 9 (1971), pp. 197–9.
24. *Yearbook of the Royal Society of London, 1907* (London, 1907), p. i.
25. E. Crawford, "The prize system of the Academy of Sciences, 1850–1914," in R. Fox and G. Weisz (eds.), *The organization of science and technology in France, 1808–1914* (Paris and Cambridge, 1980), pp. 296–8.
26. *Protokoll hållna vid sammanträden för överläggning om Alfred Nobels testamente* (Stockholm, 1899), p. 93 (Appendix B).
27. There were many other ways, not related to material needs, in which the academy benefitted from its prizes – for example, through identifying and encouraging talent that could subsequently find its place in the ranks of the academy

or using the prize system to extend the academy's influence to new areas of scientific enquiry. See Crawford, "The prize system," pp. 299–304.

28. G. Liljestrand, "The prize in physiology or medicine," in Schück et al., *Nobel: The man and his prizes*, p. 133.

29. See, for example, *prix Bordin* (1835), *prix biennal Napoleon III* (1859), *legs Maujean* (1873), *prix Baron Joest* (1880), *prix Osiris* (1899).

30. Gauja, *Les fondations de l'Académie*, pp. 316–17.

31. W. R. Nitske, *The life of Wilhelm Conrad Röntgen, discoverer of the X-ray* (Tucson, 1971), pp. 187 ff.

32. The International Bureau of Weights and Measures was proposed for the physics prize of 1908 by K. Prytz, professor at the Polytechnic Institute of Copenhagen. See *Förslag, fysik*, 1908. The abbreviations used to refer to materials in the Nobel Archives are explained in "A note on sources and notes" preceding these notes.

33. Nobel's most important donations outside his will were the gifts of 50,000 crowns each to the Karolinska Institute and the Stockholm's Children's Hospital, which he made out of the inheritance received from his mother, and his support of the arctic explorations of Nordenskiöld and Andrée. See Liljestrand, "The prize in physiology or medicine," pp. 133–4.

34. Gab [pseud.], *Monsieur Osiris* (Paris, 1911), pp. 50 ff.

CHAPTER 2 *Developments in Swedish and international science having a bearing on the Nobel institution*

1. S. Lindroth, *A history of Uppsala University, 1477–1977* (Stockholm, 1976), pp. 199–201.

2. The meaning of the Swedish term *högskola* corresponds closely to that of the German *Hochschule* – i.e., an institution of higher education that does not have the status of a university.

3. Dates of birth and death are given (in the appropriate context) only for Swedish scientists or those working in Sweden. Biographies of prominent figures in Swedish science are found in S. Lindroth (ed.), *Swedish men of science, 1650–1950* (Stockholm, 1952).

4. R. von Gizycki, "Centre and periphery in the international scientific community: Germany, France, and Great Britain in the nineteenth century," *Minerva*, 11 (1973), pp. 474–94.

5. "Fysiska Sällskapets i Uppsala protokollbok, 1887–1900," Vols. I, II, passim; "Fysiska Sällskapets i Stockholm protokollbok, 1891–1898," Vol. I, passim.

6. R. Root-Bernstein, "The Ionists: Founding physical chemistry, 1872–1890," Ph.D. dissertation, Princeton University, 1980; E. Riesenfeld, *Svante Arrhenius* (Leipzig, 1931); R. G. A. Dolby, "Debates over the theory of solution: A study of dissent in physical chemistry in the English-speaking world in the late nineteenth and early twentieth centuries," *HSPS* 7 (1976), pp. 297–404.

7. *Stockholms Högskola, 1878–1898: Berättelse över Stockholms Högskolas utveckling under hennes första tjugoårsperiod på uppdrag av hennes lärarråd utgiven av Högskolans rektor* (Stockholm, 1900), p. 81; "Främlingsbok vid Stockholms Högskola, 1893–1950," Stockholm Högskola archives, RA.

8. T. Carleman, "Magnus Gustaf Mittag-Leffler," in *Levnadsteckningar över K. Vetenskapsakademiens ledamöter*, Vol. VII (Stockholm, 1939–1948), pp. 458–71.

9. J. Sundin, *Främmande studenter vid Uppsala Universitet före andra världskriget*, Acta universitatis Upsaliensis, Skrifter rörande Uppsala universitet, C. Organisation och historia, No. 27 (Uppsala, 1973), pp. 115–18, 203 (Table 17).

10. G. Eriksson, *Kartläggarna: Naturvetenskapens tillväxt och tillämpningar i det industriella genombrottets Sverige, 1870–1914*, Acta universitatis Umensis, No. 15 (Umeå, 1978), pp. 105–154. The main theme of this study is expressed by its title (The map makers), which refers to the mapping and charting of the natural environment that the author sees as dominating Swedish science at the time.

11. Eriksson, *Kartläggarna*, pp. 19–20, 210–12.

12. Sweden and Norway had formed a dynastic union in 1815. It was dissolved peacefully in 1905, but only after some tense moments.
13. A.-L. Arrhenius-Wold, "Svante Arrhenius och utvecklingsoptimismen," in *Svante Arrhenius till 100-årsminnet av hans födelse (KVÅb 1959 Bilaga)* (Stockholm, 1959), pp. 65–100.
14. Lindroth, *A history of Uppsala University, 1477–1977*, pp. 218–20.
15. Strindberg's letters to Carlheim-Gyllensköld are reproduced in T. Eklund (ed.), *August Strindbergs brev*, Vols. VII–XIV (Stockholm, 1961–1974).
16. J. J. Thomson and H. A. Lorentz had received the Nobel prize in physics in 1906 and 1902, respectively.
17. K. Molin, "Vilhelm Carlheim-Gyllensköld som jordmagnetiker," *Kosmos*, 13 (1935), pp. 5–60.
18. The standard work on the early history of the Academy of Sciences is Lindroth, *Kungl. Svenska Vetenskapsakademiens historia, 1739–1818*. Since there is no corresponding work for the 19th and 20th centuries, the following account is based on several different sources, the most important being Eriksson, *Kartläggarna*, pp. 33–9; E. Dahlgren, *Kungl. Svenska Vetenskapsakademien: Personförteckningar, 1739–1915* (Stockholm, 1915); and the annual reports of the secretary, 1844–1903, published in the *Översikt av Kungl. Vetenskapsakademiens förhandlingar*. From 1903 onward, these reports are included in the annual yearbook of the Academy (*K. Vetenskapsakademiens årsbok*). More specialized sources are indicated by footnotes.
19. *Handlingar rörande donationer till K. Vetenskapsakademien eller under dess förvaltning* (Stockholm, 1880–1901).
20. L. Amundsen, *Det Norske Videnskapsakademi i Oslo, 1857–1957* (Oslo, 1957–1960), Vol. II, p. 93.
21. *Protokoll, KVA, 4de klassen*, April 4, 1901; *Protokoll, KVA*, May 8, 1901; S. Arrhenius, "Levnadsrön," unpublished autobiography, 1927, pp. 169–170, SUB-KVAB.
22. Letter from A. Nobel to A. E. Nordenskiöld, March 15, 1884, Nobel Collection, RA.
23. Root-Bernstein, "The Ionists," pp. 332–40.
24. B. Hasselberg, "Tobias Robert Thalén," *KVÅb 1906*, pp. 229–30; B. Hasselberg, "Om det metriska mått och viktsystemets uppkomst och utveckling," *KVÅb 1908*, pp. 199–219.
25. Lindroth, *A history of Uppsala University, 1477–1977*, p. 209.
26. Eriksson, *Kartläggarna*, pp. 14–21; A. Beckman and P. Ohlin, *Forskning och undervisning i fysik vid Uppsala universitet under fem århundraden: En kortfattad historik*, Acta universitatis upsaliensis, Skrifter rörande Uppsala universitet, C. Organisation och historia, No. 8 (Uppsala, 1965), pp. 23–6; K. Ångström, "Fysiska institutionen," in *Uppsala universitet, 1872–1897: Festskrift* (Uppsala, 1897), Vol. II, pp. 148–55.
27. Forman et al., "Physics circa 1900," pp. 30–2 refer to "unfriendly professors of experimental physics" as an obstacle to the growth of mathematical and theoretical physics. Their study also shows that, on the whole, there were very few positions officially designated as being for mathematical and theoretical physics. For example, for senior faculty there were eight in Germany, three in all of Scandinavia, and two in the United States. Official designations of positions of course only represent an approximation of the standing of the specialty in the universities.
28. I. Waller, "Carl Wilhelm Oseen," in *Levnadsteckningar*, Vol. VIII, pp. 121–43; R.M. Friedman, "Nobel physics prize in perspective," *Nature*, 292 (1981), pp. 793–8.
29. A. Fredga, "Chemistry at Uppsala until the beginning of the twentieth century," in *Faculty of Science at Uppsala University, Chemistry*, Acta universitatis upsaliensis, Uppsala University 500 years, No. 9 (Uppsala, 1976), pp. 9–12; L. Ramberg, "Oskar Widman," *KVÅb 1931*, pp. 289–334; B. Holmberg, "Olof Hammarsten," *KVÅb 1934*, pp. 272–93.

30. Arrhenius, "Levnadsrön," p. 168; Root-Bernstein, "The Ionists," pp. 22–32.
31. O. Pettersson to K. Ångström, November 10, 1883, Ångström Papers, SUB-KVAB.
32. S. Tunberg, *Stockholms Högskolas historia före 1950* (Stockholm, 1957), pp. 39–104.
33. *Stockholms Högskola, 1878–1898*, p. 77.
34. H. von Euler, "Svante Arrhenius, 1859–1927," in *Swedish men of science*, p. 232.
35. M. Pihl, "V.F.K. Bjerknes," in *DSB*, Vol. II, pp. 167–9.
36. U. Hellsten, "Ivar Fredholm, 1866–1927," in *Swedish men of science*, pp. 256–63.
37. O. Pettersson, review of *Recherches sur la conductibilité galvanique des électrolytes* by S. Arrhenius, *Nordisk Revy*, 2 (1884), pp. 204–7 (reprinted in *Svensk Kemisk Tidskrift*, 15 [1903], pp. 208–11; *Stockholms Högskola, 1878–1898*, pp. 72–103; H. von Euler, "Sven Otto Pettersson In Memoriam," *Svensk Kemisk Tidskrift*, 53 (1941), pp. 28–32; L.-O. Sundelöf and K. O. Pedersen, "Physical chemistry," in *Faculty of Science at Uppsala University, Chemistry*, p. 167.
38. Tunberg, *Stockholms Högskolas historia*, pp. 80–4.
39. Kelvin was known to be skeptical of the theory of electrolytic dissociation. The work published in 1895 by Arrhenius and N. Ekholm, dealing with lunar influences on atmospheric electricity, also went counter to Kelvin's theories, which did not admit that the earth or other planets could discharge electricity.
40. *En befordringsfråga vid Stockholms Högskola* (Stockholm, 1895), pp. 1–16.
41. It was of some solace when it was later learned that the Högskola had, in fact, figured in an earlier will (1893), but at that time only as one of several institutions, the Karolinska Institute being one of them, which *together* would have received 17 percent of Nobel's fortune. This would have put the bequest to the Högskola at half a million crowns at the most. (Tunberg, *Stockholms Högskolas historia*, pp. 86–7.)
42. See, for example, letter from O. Pettersson to N. Wille, undated (probably January 1897), Wille Papers, OUB. The polite letters that Nobel and Mittag-Leffler exchanged during their lifetimes seem to belie the assumption of a rift between the two. In 1890 for instance, Mittag-Leffler approached Nobel about the possibility of the latter endowing a chair in mathematics for Sonya Kovalevsky, who was planning to seek a position at the Academy of Sciences in St. Petersburg. Nobel politely refused the request, stating in his reply that, in his opinion, Madame Kovalevsky was better suited for Petersburg than for Stockholm. In Russia, he wrote, "the European sourdough of prejudice is reduced to a minimum" and, in any case, he did not wish for her to "be confined to a cage with her wings clipped." (Letters from G. Mittag-Leffler to A. Nobel, February 22, 1890; A. Nobel to G. Mittag-Leffler, March 1, 1890, Nobel Collection, RA.)
43. This story is probably one of the most widespread myths concerning the Nobel prize institution (see, for example, Zuckerman, *Scientific elite*, p. 18; *Science*, 215 [1981], p. 1404).
44. Eriksson, *Kartläggarna*, p. 11; *Karolinska Medikokirurgiska Institutets historia utgiven med anledning av institutets hundraårsdag* (Stockholm, 1910), Vol. I, pp. 1–286.
45. Hasselberg, "Tobias Robert Thalén," pp. 220–34; B. Hasselberg, "Untersuchungen über die Spectra der Metalle im Elektrischen Flambogen: 1–8," *K. Vetenskapsakademiens Handlingar*, 26–45 (1894–1910).
46. H. H. Hildebrandsson, "Knut Ångström," *KVÅb 1913*, pp. 303–27; M. Siegbahn, "Gustaf Granqvist," *KVÅb 1924*, pp. 308–13.
47. R. McCormmach, "Editor's foreword," *HSPS*, 3 (1971), pp. ix–xxiv.
48. The Uppsala physicists' attitudes toward theory contained elements of what Heilbron has termed "descriptionism." See J. Heilbron, "*Fin-de-siècle* physics," in Bernhard et al., *Science, technology, and society*, pp. 51–73.
49. G. Holton, *Thematic origins of scientific thought: Kepler to Einstein* (Cambridge, Mass., 1973), pp. 275–7. It is significant that Holton finds the Nobel award to Michelson "to be clearly motivated by the experimenticist philosophy of science" and quotes the presentation speech by Hasselberg at the 1907 award ceremony

to show this (*Les prix Nobel en 1907* [Stockholm, 1909], pp. 19–24). For more detail on the Nobel prize for Michelson, see the section "The specifics of consensus" in Chapter 6.

50. The view advanced by some of his biographers that Arrhenius was not a skilled experimenter may stem from his failure to meet the high standards set by the Uppsala school. Root-Berntein takes issue with this view, which in his opinion "seems to fit best Arrhenius' later work." He finds that "Arrhenius' dissertation is an irrefutable piece of evidence against ... conclusions concerning Arrhenius' experimental skill" ("The Ionists," p. 58).

51. M. Siegbahn, "Janne Rydberg, 1854–1919," in *Swedish men of science*, pp. 214–18. See also W. McGucken, *Nineteenth century spectroscopy: Development of the understanding of spectra, 1802–1897* (Baltimore and London, 1969), pp. 133–5; C. L. Maier, "The role of spectroscopy in the acceptance of an internally structured atom, 1860–1920," Ph.D. dissertation, University of Wisconsin, 1964, pp. 89–107.

52. Hasselberg, "Untersuchungen: 1," *K. Vetenskapsakademiens Handlingar*, 26, No. 5 (1894), pp. 3–4.

53. Asserting the role of a university in the capital evoked painful memories. In the early nineteenth century there had been an attempt to relocate the University of Uppsala to Stockholm. (Tunberg, *Stockholms Högskolas historia*, pp. 2–24.)

54. Pettersson was fond of neologisms. The one introduced here was derived from *Filister*, which originally was used in German to characterize "town" as opposed to "gown." Pettersson pointed out that he had borrowed from German since there was no Swedish equivalent. He hinted that the phenomenon was so common that a Swedish word would soon be found. In that he was correct, the word currently used being *fackidioti*.

55. O. Pettersson, "Svante Arrhenius," *Svensk Kemisk Tidskrift*, 15 (1903), pp. 206–7.

56. R. M. Friedman, "Constituting the polar front, 1919–1920," *Isis*, 73 (1982), p. 345.

57. E. Hiebert, "The state of physics at the turn of the century," in M. Bunge and W. R. Shea (eds.), *Rutherford and physics at the turn of the century* (New York, 1979), p. 11.

58. S. Arrhenius, *Worlds in the making* (London, 1908). This was a popular version of his *Lehrbuch der kosmischen Physik* (Leipzig, 1903). See also G. Arrhenius, "Svante Arrhenius' contribution to earth science and cosmology," in *Svante Arrhenius till 100-årsminnet av hans födelse*, pp. 101–15.

59. In von Euler's opinion, the lack of facilities and equipment "helps explain why relatively few important experimental contributions toward shaping the theory of dissociation were executed in Stockholm" ("Svante Arrhenius," p. 232).

CHAPTER 3 *Implementing the will of Alfred Nobel, 1896–1900*

1. R. Sohlman, "Alfred Nobel and the Nobel Foundation," in Schück et al., *Nobel: The man and his prizes*, pp. 37–42. This lengthy article is drawn from R. Sohlman *Ett testamente* (Stockholm, 1950; published in English under the title *The legacy of Alfred Nobel*, trans. E. H. Schubert [London, 1983]), a personal account of the settlement of Nobel's estate.

2. There was ample reason to believe that a court action in France would be crowned with success since the will did not meet the strict requirements of the Code Civil, which stipulated that only formally constituted institutions could receive bequests.

3. C. Lindhagen, *Carl Lindhagens memoarer* (Stockholm, 1936), Vol. I, pp. 287–90; S. Arrhenius, "Carl Lindhagen och Stockholms Högskola," in *Carl Lindhagen 60 år* (Stockholm, 1920), pp. 41–5.

4. Letter from C. Lindhagen to R. Sohlman, January 3, 1897, ANS-RA.

5. *Nya Dagligt Allehanda* (January 2, 1897). *Statistik över löneförhållandena inom Svenska träarbetarförbundet för år 1898* (Stockholm, 1900), pp. 17–22.
6. *Svenska Dagbladet* (January 4, 1897).
7. *Göteborgs Aftonblad* (January 4, 1897).
8. Letter from C. Lindhagen to R. Sohlman, January 16, 1897, ANS-RA.
9. "P.M. rörande Nobelska fondens användning," n.d., ANT-SA. See also Sohlman, "Alfred Nobel and the Nobel Foundation," p. 56.
10. Letter from C. Lindhagen to R. Sohlman, January 24, 1897, ANS-RA.
11. "Plan," n.d., ANT-SA; "Utkast till dispositioner i enlighet med Dr. Alfred Nobels testamente," January 1897, ANS-RA; "Alfred Nobels fond – PM angående grunder för stadgarna," March 1897, ANS-RA.
12. C. Lindhagen to R. Sohlman, January 16, 1897, ANS-RA.
13. *Protokoll, KVA*, April 14, 1897.
14. "Förklarande tillägg till Alfred Nobels testamente av den 27 November 1895 i vad det handlar om donationer för allmänna ändamål," n.d., ANT-SA.
15. *Protokoll hållna vid sammanträden för överläggning om Alfred Nobels testamente*, pp. 104–7 (Appendix G).
16. *Protokoll Alfred Nobels testamente*, pp. 99–102 (Appendix E).
17. *Nobelprotokoll, KVA*, September 15, 1900.
18. *Protokoll Alfred Nobels testamente*, pp. 98–102 (Appendices D and E).
19. Sohlman, "Alfred Nobel and the Nobel Foundation," pp. 55–6.
20. Shortly before, Forsell had failed to dissuade the Swedish Academy, of which he was also a member, from accepting responsibility for awarding the prize in literature (Sohlman, "Alfred Nobel and the Nobel Foundation," pp. 57–8; *Protokoll, KVA*, June 9, 1897; C. Lindhagen to R. Sohlman, June 11, 1897, ANS-RA).
21. Sohlman, "Alfred Nobel and the Nobel Foundation," p. 58.
22. *Protokoll Alfred Nobels testamente*, pp. 7–14; Sohlman, "Alfred Nobel and the Nobel Foundation," pp. 61–2.
23. "P. M. Den nobelska stiftelsens ändamål och grunderna för prisutdelningen," January 11, 1898, ANS-RA. The passage cited in the text was deleted from the published version of the minutes.
24. Sohlman, "Alfred Nobel and the Nobel Foundation," pp. 63–6.
25. Sohlman, *Ett testamente*, pp. 274–80.
26. Sohlman, "Carl Lindhagen och Nobelstiftelsen," in *Carl Lindhagen 60 år*, pp. 56–60.
27. O. Pettersson to C. Lindhagen, July 4, 1921, Lindhagen Collection, SA.
28. H.-G. Körber (ed.), *Aus dem wissenschaftlichen Briefwechsel Wilhelm Ostwalds* (Berlin, 1969), Vol. II, p. 151.
29. *Protokoll Alfred Nobels testamente*, pp. 115–19 (Appendix L).
30. *Protokoll Alfred Nobels testamente*, pp. 117–18 (Appendix L).
31. *Protokoll, KVA*, May 11, 1898.
32. Sohlman, *Ett testamente*, pp. 282–9.
33. C. Lindhagen to R. Sohlman, December 31, 1898, January 2, 1899, ANS-RA.
34. The last two principles had been included in the settlement with the Nobel family. The agreement also specified that prizes could not be reserved for more than five years in a row, presumably to guard against the possibility of the institutions enriching themselves by *not* awarding the prizes.
35. *Protokoll Alfred Nobels testamente*, pp. 142–52 (Appendix S).
36. *Protokoll Alfred Nobels testamente*, p. 153 (Appendix T).
37. Letter from S. Arrhenius to C. Lindhagen, January 8, 1899, ANT-SA; letter from C. Lindhagen to R. Sohlman, January 17, 1899, ANS-RA.
38. C. Salomon-Bayet, "Bacteriology and Nobel prize selections, 1901–1920," in Bernhard et al. (eds.), *Science, technology, and society*, p. 376.
39. Liljestrand, "The prize in physiology or medicine," pp. 135–9.
40. *Protokoll Alfred Nobels testamente*, pp. 44–6.
41. S. Arrhenius to C. Lindhagen, February 15, 1899, Lindhagen Collection, SA.

42. Lindhagen, *Carl Lindhagen's memoarer*, p. 278.
43. *Protokoll, KVA*, May 23, June 7, 1899.
44. Konseljprotokoll, ecklesiastikärenden, June 29, 1900, RA.

CHAPTER 4 *An overview of the nominating system and its influence on the prize decisions*

1. *Nobelprotokoll, KVA*, September 15, 1900. For the list of members elected in subsequent years, see Appendix C.
2. Since the categories of permanent nominators had been defined in the special regulations (para. 1.1–4), the committee decided on the two categories (chairholders at foreign universities and scientists invited in their individual capacity) to whom invitations were sent each year in accordance with para. 1.5–6. In the first year, the invitation was extended to chairholders in both physics and chemistry at nine universities, including the University of Berlin and the Science Faculty at the Sorbonne. From 1902 onward, the physics and chemistry committees drew up separate lists of the universities where chairholders were invited. Invitations to scientists to nominate in their individual capacity were not extended until 1903 in physics and 1904 in chemistry.
3. The motives that made the nominators prefer a particular candidate can only be discerned, of course, in a few well-documented cases, principally those where the nominations resulted from organized campaigns. For this reason, the term *vote* is not to be taken literally here and in the following: Letters of nomination bore little resemblance to "votes" since at the very least they would give the name of a candidate and mention the work for which he was proposed; sometimes they would extend to lengthy treatises on the candidate's entire career.
4. For a candidate to be recommended for the prize, it was sufficient for his or her name to be put forward by *one* nominator.
5. *KU, fysik*, 1903.
6. A nomination is counted as such each time a candidate's name is put forward, irrespective of whether he or she was proposed for a divided prize or figures as an alternative candidate in the proposal.

   Nominations were considered invalid (1) if nominators proposed themselves for the prizes; (2) if nominations were of deceased scientists; (3) if nominations were received after February 1; (4) if nominations were made for a scientist's life's work rather than for a specific discovery or invention. On points 3 and 4 the physics committee was more lenient than the chemistry (cf. Chapter 6); this explains why a larger number of nominations were disallowed in chemistry than in physics (sixty-one as compared with twenty-three).
7. In case of divided prizes, Figure 4.1 shows the number for the *individual* prize-winner receiving the most nominations in a given year or cumulatively.
8. Nernst was awarded the prize for 1920 in 1921.
9. Among the *Swedish* members of the Academy, the most active nominators were those who served on the Nobel Committees for Physics and Chemistry and whose nominations, therefore, figure in category 2. An insignificant portion of nominations (less than 5 percent) came from other Swedish members of the Academy.
10. *Protokoll, NK, fysik*, January 8, 1901.
11. See, for example, the statement of Arrhenius in the minutes of the physics committee (*Protokoll, NK, fysik*, August 16, 1902) where he explains that he had proposed Zeeman for the prize of 1902 only to make sure that Zeeman would not be excluded on formal grounds (i.e., lack of a proposal) from being considered for the prize.
12. H. Ramser, "Warburg, Emil Gabriel" *DSB*, Vol. XIV, pp. 170–2. The following quotation provides some insight into Warburg's disposition: "Of Warburg's approximately 150 publications only one is of a polemical nature; even then he did not begin the dispute, and it did not concern any scientific matter. All his

other writings display a sober objectivity, and critical detachment from his own results" (p. 172).

13. See, for example, interview with H. Urey (Chemistry, 1934) in the *New York Times* (December 7, 1975).

14. Nominators, candidates, and prizewinners figure in the tables under the name of the country in which they were working at the time and *not* by the nationality they held at that time. Actually, in most cases the person was working in the country of which he or she was also a national.

15. In the period 1901–1915, German scholars constituted one-third of the foreign membership of the Academy of Sciences.

16. "*Chauvinism*" has been put in quotes here to indicate that nominators' preferences for their countrymen could reflect considerations other than strictly nationality. Among such other considerations one might mention linguistic abilities or the scientific styles that nominators of the same nationality had in common.

17. American candidates in chemistry received eight nominations from British nominators as compared with seven from German ones. If the German nominators are grouped with those from Central and Eastern Europe, the latter figure rises to thirteen. In the case of candidates in other Nordic countries, the number of nominations received from Central and Eastern Europe was also very low (four); furthermore, they all emanated from one nominator (M. Bamberger in Vienna).

CHAPTER 5 *Networks at work in the prize selections: Arrhenius and Mittag-Leffler*

1. The different ways in which Arrhenius and Mittg-Leffler could bring their influence to bear on Nobel prize selections are reflected in their correspondence. Arrhenius, as befitted his position as a member of the Nobel Committee for Physics, was notably discreet in writing to foreign colleagues about matters involving the prizes (or – but this is unlikely – letters bearing upon such matters were destroyed by their recipients). By contrast, the support that Mittag-Leffler could build for his campaigns depended not only on his presenting his correspondents with a "battle plan" but also on keeping them informed about the unfolding of events at different stages of the decision-making process.

2. Körber, *Aus dem Briefwechsel Wilhelm Ostwalds*, Vol. II, p. 194.

3. O. Arrhenius, "Svante Arrhenius – det första kvartsseklet," in *Svante Arrhenius till 100-årsminnet av hans födelse*, p. 63; Root-Bernstein, "The Ionists," p. 114.

4. It is not correct, as Arrhenius stated in the autobiographical notice he wrote on the occasion of his Nobel prize, that "the dissertation received only one grade above a failing one," *non sine laude approbatur* being, in fact, two grades above a fail (*Les prix Nobel en 1903* [Stockholm, 1906], p.67). That this version of the event remained a thorn in the side of Uppsala scientists is illustrated by the reply of K. Ångström to Mittag-Leffler, who had inquired about the veracity of the statement quoted above. Ångström ends his letter as follows: "Poor Cleve [member of the jury and chairman of the Nobel Committee for Chemistry]! He tried to repair his unforgiveable mistake through a Nobel prize – and nobody championed this more than he – but for many more years to come the story of the idiotic Uppsala faculty will be told" (K. Ångström to G. Mittag-Leffler, March 2, 1909, Mittag-Leffler Collection, M.-L. Inst.).

5. S. Arrhenius to V. Bjerknes, June 11, 1904, Bjerknes Collection, OUB.

6. Arrhenius, "Levnadsrön," p. 142.

7. Arrhenius's father administered the land holdings of Uppsala University and Mittag-Leffler's was headmaster of a Stockholm secondary school. There was no wealth in either family, but Mittag-Leffler was active in numerous business ventures that involved, among other things, the rapidly expanding insurance industry.

8. O. Pettersson to V. Bjerknes, November 15, 1907, Bjerknes Collection, OUB.

9. A great number of letters were exchanged in 1898–1900 between Mittag-Leffler

and his French correspondents (G. Darboux, P. Painlevé, E. Picard, and H. Poincaré) on the subject of his election to the Paris Academy (Mittag-Leffler Collection, M.-L. Inst.).

10. G. Mittag-Leffler to P. Painlevé, July 18, 1902, Mittag-Leffler Collection, M.-L. Inst.

11. Carleman, "Magnus Gustaf Mittag-Leffler," p. 469.

12. H. Poincaré, "Analyse des travaux scientifiques de Henri Poincaré faite par lui-même," *Acta Mathematica*, 38 (1921), p. 130 (emphasis added). See also L. Pyenson, "Relativity in late Wilhelmian Germany: The appeal to a preestablished harmony between mathematics and physics," *Archive for History of Exact Sciences*, 27 (1982), p. 140.

13. Darboux was permanent secretary representing *sciences mathématiques* from 1900 to 1917, while M. Berthelot held the same position for *sciences physiques* from 1889 to 1907.

14. In physics, corporative nominations were made in favor of E. H. Amagat, H. Becquerel, P. Curie, G. Lippmann, and H. Poincaré; in chemistry, only the candidacy of H. Moissan was supported in this manner.

15. S. Arrhenius to C. Aurivillius, June 12, 1909, Aurivillius Papers, SUB-KVAB.

16. S. Arrhenius, *Theories of solutions*, Silliman Memorial Lectures (New Haven, 1912).

17. That the peregrinations of Arrhenius and Mittag-Leffler irritated some of their colleagues at home is illustrated by the following comment in a letter by Hasselberg: "Our new travelling salesman of science, Svante, has not yet returned from his business trip.... However, Mittag-Leffler 'le grand commis voyageur des mathématiques' has just returned from Rome" (B. Hasselberg to K. Ångström, January 1, 1904, Ångström Papers, SUB-KVAB).

18. C. W. Oseen, *Atomistiska föreställningar i nutidens fysik: Tid, rum, och materia* (Stockholm, 1919), pp. 9–10.

19. S. Arrhenius, "Electrolytic dissociation versus hydration," *Philosophical Magazine*, 28 (1889), pp. 30–1; A. Ölander, "Arrhenius och den elektrolytiska dissociationsteorin," in *Svante Arrhenius till 100-årsminnet av hans födelse*, pp. 17–21.

20. Dolby, "Debates over the theory of solution," p. 391.

21. *Förslag, kemi*, 1901.

22. J. R. Partington, *A history of chemistry* (London, 1972), Vol. IV, pp. 675–7; Root-Bernstein, "The Ionists," pp. 114–19, 375–86.

23. There were also problems of simultaneous discovery in that Arrhenius and Planck had announced similar versions of the dissociation theory in 1887, but these were not touched on in any of the committee reports (Partington, *A history of chemistry*, Vol. IV, p. 679; Root-Bernstein, "The Ionists," pp. 460–73).

24. Nobel Foundation, *Nobel lectures: Chemistry, 1901–1921* (Amsterdam, 1967), p. 9; Partington, *A history of chemistry*, Vol. IV, p. 655; Root-Bernstein "The Ionists," pp. 482–5.

25. *Protokoll, NK, fysik*, August 16, 1902, August 5, 1903; *Nobelprotokoll, KVA*, September 9, 1903.

26. *Förslag, fysik, kemi*, 1903.

27. *Protokoll, NK, kemi*, April 5, 190? For the text of the letter, see *Protokoll, NK, fysik*, May 9, 1903 (Appendix 2). See also J. W. S. Rayleigh and W. Ramsay, "Argon, a new constituent of the atmosphere," *Philosophical Transactions of the Royal Society* (A) 186 (1896), pp. 187–241; E. Hiebert, "Historical remarks on the discovery of argon: The first noble gas," in H. H. Hyman (ed.), *Noble-gas compounds* (Chicago, 1963), pp. 3–20.

28. For the text of the letter, see *Protokoll, NK, kemi*, May 20, 1903 (Appendix).

29. O. Widman to H. Söderbaum, August 26, 1903, Söderbaum Collection, SUB-KVAB.

30. *Protokoll, NK, kemi*, May 20, 1903. The different types of committee reports are described in "A note on sources and notes" preceding the notes. The role of the reports in committee decision making is discussed in Chapter 6.

31. *KVÅb, 1904*, p. 38. In a letter to the secretary of the Academy, Ångström insisted that the lecture be held *before* the plenary meeting at which the Academy decided on the year's prizewinners (K. Ångström to C. Aurivillius, October 11, 1903, Aurivillius Papers, SUB-KVAB).

32. *Nobelprotokoll, KVA, 4de klassen*, October 31, 1903 (statement by Cleve); *KU, fysik*, 1903; *Nobelprotokoll, KVA*, November 12, 1903.

33. *Protokoll, NK, kemi*, September 19, 1903; *Nobelprotokoll, KVA, 5te klassen*, October 31, 1903.

34. "Motiverade vota för Nobelpris, m.m.," Widman Collection, UUB; *Nobelprotokoll, KVA*, November 12, 1903.

35. *Nobelprotokoll, KVA*, November 12, 1903.

36. *KU, kemi*, 1904.

37. E. Hiebert and H.-G. Körber, "Ostwald, Friedrich Wilhelm," in *DSB*, Vol. XV, Suppl. 1, p. 456.

38. *Zeitschrift für physikalische Chemie*, 9 (1892), p. 771. See also Hiebert and Körber, "Ostwald, F. W.," pp. 462–4; N. Holt, "A note on Wilhelm Ostwald's energism," *Isis*, 61 (1970), pp. 386–9.

39. *Aftonbladet* (February 11, 1896).

40. S. Arrhenius, "Atomlärans framgångar," *Svensk Kemisk Tidskrift*, 20 (1908), pp. 173–4; S. Arrhenius to J. Loeb, September 3, 1908, Loeb Collection, LC. Ostwald's conversion was first recorded in his review of T. Svedberg's *Studien zur Lehre von den kolloiden Lösungen* in *Zeitschrift für physikalische Chemie*, 64 (1908), pp. 508–9. A more elaborate statement appeared in the preface to the fourth (revised) edition of his *Grundriss der allgemeinen Chemie* (1909).

41. *Förslag, kemi*, 1909.

42. S. Arrhenius to O. Widman, April 1, 1909, Widman Collection, UUB.

43. O. Widman to S. Arrhenius, April 28, 1909, Arrhenius Collection, SUB-KVAB.

44. O. Widman to S. Arrhenius, June 9, 1909, Arrhenius Collection, SUB-KVAB.

45. *KU, kemi*, 1909. See also S. Arrhenius to O. Widman, July 7, 13, 1909, Widman Collection, UUB; O. Widman to S. Arrhenius, July 12, 1909, Arrhenius Collection, SUB-KVAB.

46. Partington, *A history of chemistry*, Vol. IV, p. 600; Hiebert and Körber, "Ostwald, F. W.," p. 462.

47. E. Hiebert, "Nernst, Hermann Walther," in *DSB*, Vol. XV, Suppl. 1, pp. 433–4.

48. That the reconciliation between Arrhenius and Nernst was a sine qua non for the award of the Nobel prize to the latter is brought out in letters – for example, H. G. Söderbaum to O. Widman, August 3, 1921, Widman Collection, UUB; G. Bredig to S. Arrhenius, February 17, 1922, Arrhenius Collection, SUB-KVAB.

49. A. Lundgren, "Arrhenius om van't Hoff och Ehrlich," *Lychnos* (1975–1976), p. 94; Riesenfeld, *Svante Arrhenius*, p. 53.

50. "While Ostwald unquestionably was the most influential *initial* spokesman and organizer of the new school of physical chemistry, his numerous textbooks were devoted to general, inorganic, and analytical chemistry and did not serve the function of a useful repository for theoretical or physical chemistry. It was Nernst's *Theoretische Chemie* [1st ed., 1893] that set the pattern for the 20th century along this line" (E. Hiebert, "Nernst and electrochemistry," in G. Dubpernell and J. H. Westbrook [eds.], *Selected topics in the history of electrochemistry* [Princeton, 1978], p. 187).

51. Nernst used some of the proceeds to acquire a hotel and café – known as Café Nernst – in Göttingen (K. Mendelssohn, *The world of Walther Nernst: The rise and fall of German science* [London, 1973], pp. 44–9). When G. Tamman nominated Nernst for the chemistry prize of 1910, Arrhenius, who was usually discreet when it came to Nobel prize matters, launched into a diatribe against Nernst, accusing him of having "made" rather than "earned" money on his lamp and of having invested his money in "Nachtcaféen" where it earned high interest.

"People here think," he says, "that there are enough 'Nachtcaféen' without these having to be supported with Nobel money." He goes on to say: "Unfortunately morals and science do not always coexist." Tamman answered with a lengthy defense of Nernst's morals. (S. Arrhenius to G. Tamman, December 22, 1910; G. Tamman to S. Arrhenius, December 24, 1910, Arrhenius Collection, SUB-KVAB.)

52. *Utkast, KU, kemi,* June 4, 1907; September 5, 6, 1908; August 30, 1910; May 26, 1914.

53. For a more detailed account of Arrhenius's interest in atomism, see E. Crawford, "Arrhenius, the atomic hypothesis, and the 1908 Nobel prizes in physics and chemistry," *Isis,* 75 (1984), forthcoming.

54. Post has defined as "essentially atomic" those theories or lines of investigation that allowed determinations of *N*, that is, the number of molecules in a gram-molecule of any substance (Avogadro's number). (H. R. Post, "Atomism 1900, I and II," *Physics Education,* 3 [1968], pp. 226–7.)

55. If *e* was known, *N* could be determined, and vice versa. For an overview of different methods of determining *e* and *N*, see D. Anderson, *The discovery of the electron: The development of the atomic concept of electricity* (Princeton, 1964), pp. 74–87. The specific determinations mentioned here are discussed in M. J. Nye, *Molecular reality: A perspective on the scientific work of Jean Perrin* (London and New York, 1972), pp. 97–142; L. Badash, "Rutherford, Ernest," in *DSB,* Vol. XII, p. 30; T. S. Kuhn, *Black-body theory and the quantum discontinuity, 1894–1912* (New York, 1978), pp. 111–12. For general developments in atomic theory, see, D. M. Knight, *Atoms and elements: A study of theories of matter in England in the nineteenth century* (London, 1967); A. G. van Melsen, *From atmos to atom* (Pittsburgh, 1952); B. Schonland, *The atomists (1805–1933)* (Oxford, 1968).

56. See, for example, Ramsay's presidential address delivered at the annual meeting of the Chemical Society (W. Ramsay, "The electron as an element," *Journal of the Chemical Society: Transactions,* 93 [1908], pp. 774–88).

57. *Förslag, fysik, kemi,* 1908.

58. T. Trenn, *The self-splitting atom: The history of the Rutherford-Soddy collaboration* (London, 1977), pp. 144–5.

59. N. Feather, "Some episodes of the alpha-particle story, 1903–1977," in Bunge and Shea, *Rutherford and physics at the turn of the century,* pp. 74–5.

60. *KU, fysik,* 1906.

61. *Protokoll, NK, kemi,* May 2, 1908.

62. *Utkast, KU, kemi,* September 5, 6, 1908.

63. For a detailed account of materials in the archives of the Nobel Committee for Physics relating to the Nobel prize for Planck during the period from 1907 (the year Planck was nominated for the first time) to 1919 (when Planck received the prize reserved in 1918), see B. Nagel, "The discussion concerning the Nobel prize for Max Planck," in Bernhard et al., *Science, technology, and society,* pp. 352–76.

64. Kuhn, *Black-body theory,* pp. 92–135; Nagel, "The discussion concerning the Nobel prize for Max Planck," pp. 354–5.

65. *KU, fysik,* 1908.

66. *Protokoll, NK, fysik,* September 18, 1908 (Appendix C).

67. *Nobelprotokoll, KVA,* November 10, 1908; *Nobelprotokoll, KVA, 3dje klassen,* October 31, 1908. According to Mittag-Leffler, thirteen votes were cast for Planck as against forty-six for Lippmann (G. Mittag-Leffler to P. Painlevé, December 9, 1908, Mittag-Leffler Collection, M.-L. Inst.).

68. Nagel, "The discussion concerning the Nobel prize for Max Planck," pp. 363–4.

69. In the lecture to the Society of Chemists where Arrhenius announced Ostwald's conversion to an atomist viewpoint, he first discussed Rutherford's and Planck's determinations of *e* at some length, and then went on: "Planck's theory is also of interest because it assumes that energy, too, has an atomistic quality in that

there exists a minimal value for energy below which one cannot move" ("Atom-lärans framgångar," p. 117). It should be noted that the lecture was given *after* the plenary meeting in the Academy.

70. *Nobelprotokoll, KVA, 3dje klassen*, October 31, 1908.

71. Kuhn, *Black-body theory*, pp. 189–96.

72. The document marked "Fredholm, November 10, 1908," Mittag-Leffler Collection, M.-L. Inst. Nagel's account (p. 364) sheds light on this document.

73. G. Mittag-Leffler, "Diaries," November 10, 1908, Mittag-Leffler Papers, KB. Wien, who went down to defeat in the Academy, prophetically put his finger on the matter of Mittag-Leffler's intervention when he wrote to A. Sommerfeld expressing disappointment over Lorentz's lecture and added: "What is the purpose of presenting this question to the mathematicians, none of whom is equipped to make a judgment on this sort of point." (W. Wien to A. Sommerfeld, June 6, 1908, quoted in Kuhn, *Black-body theory*, p. 192).

74. N. Ekholm to S. Arrhenius, March 10, 1910, Arrhenius Collection, SUB-KVAB. Receiving the news about the award, Rutherford wrote to Arrhenius: "I am naturally very pleased at the award though I feel very unworthy to be included in the chemical list of prizewinners" (E. Rutherford to S. Arrhenius, November 17, 1908, Arrhenius Collection, SUB-KVAB).

75. *Leipzig Neueste Nachrichten* (November 23, 1908); *Nature*, 79 (December 3, 1908), p. 138. See also, for example, *Berlin Lokal Anzeiger* (November 24, 1908); *Daily Mail* (November 24, 1908); *Le Figaro* (November 21, 1908); *Die Zeit* (November 24, 1908).

76. Kuhn, *Black-body theory*, pp. 193–4.

77. Planck received the 1918 physics prize in 1919. See, Nagel, "The discussion concerning the Nobel prize for Max Planck," pp. 364–74.

78. Hiebert, "Nernst, W.H.," in *DSB*, pp. 442–3; Kuhn, *Black-body theory*, pp. 210–20.

79. See, for example, *KU, fysik*, 1912. That recognition of Planck's theory by a Nobel award was strongly linked in Arrhenius's mind to one for Nernst is brought out in one of his letters to T. W. Richards. After having announced that both J. J. Thomson and J. H. Jeans had now been won over to the theory of quanta, he stated: "I think that the theory will soon be generally accepted.... You know that Nernst has been a good politician in connecting his third theorem with the theory of quanta ... and that this circumstance makes that he will float with Planck. I am quite of your opinion that Nernst's merit is not very great in this point, but the public, especially in Germany, will give him the credit when Planck is recognized" (S. Arrhenius to T. W. Richards, November 25, 1913, Richards Papers, Harvard University Library).

80. A. Westgren, "The chemistry prize," in Schück et al., *Nobel: The man and his prizes*, p. 406.

81. *Förslag, fysik, kemi*, 1915. Arrhenius had nominated Moseley for the prizes in both physics and chemistry, citing his articles "The high frequency spectra of the elements: Parts I and II" (*Philosophical Magazine*, 26 [1913], pp. 1024–34; 27 [1914], pp. 703–13). Probably at the instigation of Arrhenius, his candidacy was assigned, in a joint meeting of the committees, to the chemistry committee. See also J. Heilbron, *H. G. J. Moseley: The life and letters of an English physicist, 1887–1915* (Berkeley, Calif., 1974).

82. *KU, kemi*, 1915.

83. R. McCormmach, "Lorentz, H. A.," *DSB*, Vol. IV, pp. 487–500; T. Hirosige, "Electrodynamics before the theory of relativity, 1890–1905"; *Japanese Studies in the History of Science*, 5 (1966), 1–49.

84. G. Mittag-Leffler to P. Painlevé, August 29, 1902. See also G. Mittag-Leffler to H. Poincaré, February 5, 1902. All letters by and to Mittag-Leffler cited here and below are in the Mittag-Leffler Collection, M.-L. Inst.

85. G. Mittag-Leffler to H. Poincaré, December 14, 1901.

86. H. Poincaré, "Relations entre la physique expérimentale et la physique mathé-

matique," in C. E. Guillaume and L. Poincaré (eds.), *Rapports présentés au Congrès International de Physique, Paris 1900* (Paris, 1900–1901), Vol. I, pp. 1–29.

87. *Förslag, fysik*, 1902.
88. E. Whittaker, *A history of theories of aether and electricity* (London, 1953), Vol. I, pp. 413–414.
89. H. Poincaré, *Electricité et optique: La lumière et les théories éléctrodynamiques*, Leçons professées à la Sorbonne en 1888, 1890, et 1899 (Paris, 1901), p. 573.
90. *Förslag, fysik*, 1902.
91. *Protokoll, NK, fysik*, August 16, 1902.
92. G. Mittag-Leffler to H. A. Lorentz, November 12, 1902.
93. Nobel Foundation, *Nobel lectures: Physics, 1901–1921* (Amsterdam, 1967), p. 27.
94. See Chapter 4, note 11.
95. *KU, fysik*, 1902. More recently Anderson has stated: "The fact that Zeeman's early observations did not show forth this multiple splitting provides one of the occasional cases in physics in which the preliminary absence of thoroughly precise, fine-grained data aided rather than hindered the development of a useful model or theory" (Anderson, *The discovery of the electron*, p. 49).
96. G. Mittag-Leffler to H. A. Lorentz, November 12, 1902; G. Mittag-Leffler to P. Painlevé, August 29, 1902.
97. G. Mittag-Leffler, "Diaries," November 12, 1903, Mittag-Leffler Papers, KB.
98. *Förslag, fysik*, 1903.
99. For Marie Curie's ill-fated attempt at being elected in 1911, see R. Reid, *Marie Curie* (London, 1974), pp. 178–88.
100. P. Curie to G. Mittag-Leffler, August 6, 1903. Mittag-Leffler's earlier letter to Pierre Curie has not been found either in the Mittag-Leffler Collection or in the Curie Collection at the Bibliothèque Nationale in Paris.
101. *Protokoll, NK, fysik*, August 29, 1903.
102. G. Mittag-Leffler to P. Painlevé, December 9, 1908.
103. The year had opened with H. Farman, in a Voisin machine, accomplishing the first circling maneuver by an aircraft during a flight that lasted for 88 seconds. In March a syndicate, the Compagnie Générale de Navigation Aérienne, was formed in order to build, sell, or license the Wright airplane in France, and in May O. Wright arrived in Le Havre to carry out the tests required by the contract he and his brother had signed with the syndicate. Flying his plane at Le Mans throughout the autumn, Wright, alone or with passengers, steadily improved his performance, setting new records for height, distance, and maneuverability. Meanwhile Farman, with Painlevé on board, set the French distance record for a passenger flight and became the first to fly between two cities: Châlons and Reims. To obtain political support and funds, November 5, 1908 was declared the first official Day of Aviation and marked by ceremonies and speeches in Parliament. Speaking on the occasion, Painlevé concluded that "no effort will be spared to ensure that France remains what she is today, that is, the center of aviation" (J. D'Estournelles de Constant, P. Painlevé, and V.-P. Bouttieaux, *Pour l'aviation* [Paris, 1909], p. 122). See also H. Combs, *Kill Devil Hill: Discovering the secrets of the Wright brothers* (Boston, 1979), pp. 278–316.
104. P. Painlevé to G. Mittag-Leffler, December 28, 1908.
105. P. Painlevé to G. Mittag-Leffler, January (n.d.), 1909.
106. G. Mittag-Leffler to P. Painlevé, February 3 (two letters), October 27, 1909.
107. *Förslag, fysik*, 1909.
108. *KU, fysik*, 1909. The committee's fears were not unwarranted, for after the successes of 1908 there had been a rapid rise in accidents involving airplanes, with a total of eleven deaths recorded between late 1908 and mid-1910. The reviewer of a spate of books on aeronautics in *Nature* who cited these figures was thus prompted to observe that "it would have been better, cheaper and probably quite as quick in the long run to have first got everything done that could be done in studying the problems of aviation by the methods of exact science and to have developed the practical side subsequently" (*Nature*, 84 [1910], p. 229).

109. G. Mittag-Leffler, "Diaries," November 9, 1909, Mittag-Leffler Papers, KB.
110. G. Mittag-Leffler to P. Appell, November 28, 1909.
111. Copies of the large number of letters sent to scientists around the world are found at the Mittag-Leffler Institute. See also G. Mittag-Leffler to P. Painlevé, November 28, 1909; idem to P. Appell, December 17, 1909; idem to G. Darboux, January 3, 1910; G. Mittag-Leffler, "Diaries," November 18, December 17, 1909, Mittag-Leffler Papers, KB.
112. The two other French laureates, Becquerel and Pierre Curie, died in 1908 and 1906, respectively.
113. A. Schuster to G. Mittag-Leffler, December 27, 1909.
114. E. Rutherford to G. Mittag-Leffler, January 16, 1910.
115. J. J. Thomson to S. Arrhenius, January 7, 1909 (probably should be 1910), Arrhenius Collection, SUB-KVAB.
116. *KU, fysik*, 1909.
117. G. Mittag-Leffler to B. Hasselberg and V. Carlheim-Gyllensköld, April 27, 1910.
118. *KU, fysik*, 1910.
119. *Förslag, fysik*, 1910.
120. B. Hasselberg to G. Mittag-Leffler, January 24, 1910.
121. *Protokoll, KVA, 3dje klassen*, April 2, 1910; *Nobelprotokoll, KVA*, April 13, 1910.
122. *Förslag, fysik*, 1910.
123. *KU, fysik*, 1910.
124. *Protokoll, NK, fysik*, September 26, 1910 (the dissenting opinion of Hasselberg and Carlheim-Gyllensköld is included as Appendix 4); *Nobelprotokoll, KVA, 3dje klassen*, October 22, 1910; *Nobelprotokoll, KVA*, November 5, 1910; G. Mittag-Leffler to G. Darboux, February 25, 1911.
125. *Förslag, fysik*, 1911. The special report on Poincaré that Carlheim-Gyllensköld wrote for the committee (*Utkast, KU, fysik*, September 2, 1911) also figures as his dissenting opinion (*Protokoll, NK, fysik*, September 20, 1911).
126. *Protokoll, NK, fysik*, September 20, October 20, 1911.
127. J. Torbacke, "Affären Staaff – Mittag-Leffler," *Statsvetenskaplig Tidskrift*, 1 (1961), pp. 10–42.
128. G. Mittag-Leffler, "Diaries," July 19, 1922, Mittag-Leffler Papers, KB.

CHAPTER 6 *Committee decision making*

1. *Nobelprotokoll, KVA*, October 19, 1901.
2. *Nobelprotokoll, KVA*, November 5, 1902.
3. *Protokoll, gem. sam.*, February 11, 1903.
4. *Nobelprotokoll, KVA*, April 25, 1903.
5. *Nobelprotokoll, KVA, 3dje och 4de klasser*, November 12, 1904; *Nobelprotokoll, KVA*, November 26, 1904.
6. *Nobelprotokoll, KVA, 4de klassen*, December 6, 1904; *Nobelprotokoll, KVA, 3dje klassen*, January 30, 1905; *Nobelprotokoll, KVA*, January 21 and March 1, 1905 (the accusations and counteraccusations presented by Hasselberg and Arrhenius are included as appendices). See also the file labeled "Arrhenius–Hasselbergska kontroversen, 1904–1905," Hasselberg Papers, UUB; "Promemoria" accompanying letter from S. Arrhenius to G. Retzius, December 21, 1904, Retzius Collection, SUB-KVAB; S. Arrhenius to J. H. van't Hoff, December 1 and 18, 1904, and March 2, 1905, van't Hoff Papers, Johns Hopkins University Library, Baltimore.
7. The 1,000 crowns paid for the sample (prepared under the personal supervision of Mme. Curie) was well below the price on the open market; in 1903, the quote from the Giesel firm – the main source of commercially prepared radium – would have been 2,750 crowns. When Giesel discontinued sales in 1906, it had risen to 11,000 crowns, an increase due both to scientists' interest in radioactivity studies and to the possibilities of application in medicine. It is ironic that the Academy, whose award of the Nobel prizes to the Curies contributed so much to this

enhanced interest, could not in 1904 purchase the small additional quantity of radium needed to conclude Ångström's investigations, since the price had risen too rapidly. However, Marie Curie herself experienced similar difficulties and had to devote much energy to private fundraising to secure her working materials. (*Protokoll, NK, fysik*, April 18 and August 29, 1903, and January 31, 1904; Forman et al., "Physics *circa* 1900," p. 89; Reid, *Marie Curie*, pp. 141–4, 257–70.)

8. K. Ångström, "Contributions à la connaissance du dégagement de chaleur du radium," *Arkiv för matematik, astronomi, och fysik*, 1–2, (1904–1905), pp. 1–7, 523–8. See the section "The influence of Arrhenius" in Chapter 5.

9. See, e.g., C. Kirsten and H.G. Körber (eds.), *Physiker über Physiker: Wahlvorschläge zur Aufnahme von Physikern in die Berliner Akademie 1870 bis 1929* (Berlin, 1975).

10. *Protokoll, NK, kemi*, March 30, 1901; *Protokoll, NK, fysik*, May 7, 1901; *Nobelprotokoll, KVA*, September 21, 1901.

11. *Protokoll, gem. sam.*, March 28, 1908, and April 15, 1908.

12. *Protokoll, gem. sam.*, December 1, 1900.

13. *Protokoll, gem. sam.*, January 10, 1907; *Nobelprotokoll, KVA*, January 26, 1907.

14. *Nobelprotokoll, KVA*, December 18, 1907.

15. *Protokoll, gem. sam.*, April 15, 1908.

16. Ziman gives the general meaning of this term when he writes: "until the provocative, unexpected message has been to some extent 'processed' by the social instrument of the scientific community it can neither be ignored nor taken as a basis for action, and thus lies outside the dimensions of scientific belief and doubt" (J. Ziman, *Reliable knowledge* [Cambridge, 1978], pp. 139–40).

17. Reference to G. Le Bon and R. Blondlot are found in *KU, fysik*, 1903 and 1905, respectively. For analyses of these episodes with reference to the French intellectual climate and scientific milieu of the time, see M. J. Nye, "Gustave Le Bon's black light: A study in physics and philosophy in France at the turn of the century," *HSPS*, 4 (1974), pp. 163–95; idem, "N-rays: An episode in the history and psychology of science," *HSPS*, 11 (1980), pp. 125–56. Although the French Academy of Sciences had awarded Blondlot the *prix Leconte* in the amount of 50,000 francs in 1904, no influential member of the Academy supported him for the Nobel prize, probably because of the controversy that surrounded the N-rays shortly afterward.

18. *KU, fysik*, 1908, 1911, 1913, 1915.

19. *Protokoll, NK, fysik*, February 2, 1901. In the negotiations over the statutes, Sohlman had proposed that patented inventions be the object of prize awards only if their authors relinquish patent rights. In the Large Nobel Committee, an equal number of votes were cast for and against this provision, and it was not included in the statutes (*Protokoll hållna vid sammanträden för överläggning om Alfred Nobels testamente*, pp. 29–30).

20. *KU, kemi*, 1905.

21. *Förslag, fysik*, 1901 (letter from E. Warburg). That both committees successfully fought the inclination on the part of the nominators to propose older scientists (and works) for the prizes is demonstrated by the fact that whereas slightly more than half of the nominations in physics and somewhat less than half of those in chemistry were of scientists who were fifty-five years or older, only about one-third of the laureates in both disciplines fell into this age bracket.

22. S. Arrhenius to N. Wille, November 17, 1908, Wille Collection, OUB.

23. *KU, kemi*, 1908. At the suggestion of Pettersson, Ramsay wrote to Arrhenius asking him to intervene in favor of Crookes. Ramsay's chief argument was that "it is hard that others, who may be regarded as in a sense Crookes's pupils, should be rewarded and yet he goes without recognition." (W. Ramsay to S. Arrhenius, October 1, 1907, Arrhenius Collection, SUB-KVAB.)

24. *KU, kemi*, 1906; *Nobelprotokoll, KVA, 4de klassen*, October 27, 1906.

25. The linking of van der Waals and Kamerlingh Onnes is particularly apparent in

the general report of 1909. See also J. van den Handel, "Kamerlingh Onnes, Heike," *DSB*, Vol. VII, pp. 220–2.

26. *KU, fysik*, 1903 and 1905. The report of the latter year reveals a more positive attitude toward the eventual rewarding of Boltzmann.

27. *Nobelprotokoll, KVA*, November 12, 1901.

28. E. Phragmén to G. Mittag-Leffler, November 19, 1901, Mittag-Leffler Collection, M.-L. Inst.

29. A. Ihde, *The development of modern chemistry* (New York, 1964), pp. 458, 634, 682. Moissan's claim – which the committee wisely did not include in the award citation – created a sensation at the time but was later disproved (F. P. Bundy, H. T. Hall, H. M. Strong, and R. H. Wentorf, "Man-made diamonds," *Nature*, 156 [1955], pp. 51–5).

30. *KU, fysik*, 1908.

31. For an account of the circumstances leading up to the "revolt" in the Academy in 1912, see Eriksson, *Kartläggarna*, pp. 95–7.

32. Overall, this process fits the concept of "bounded rationality" elaborated by Simon (H. Simon, *Administrative behaviour* [London and New York, 1976], pp. 80–3, 240–4). In Simon's opinion, no decision maker conforms to the model of a "rational man" who, before arriving at a decision, weighs all available alternatives as well as the whole complex of consequences that would follow from each choice. Decision making takes place, rather, within an area of rationality "bounded" by the irrational and nonrational elements that more often than not determine behavior. The need to arrive at consensus was the main mechanism for setting limits to nonrational elements in decisions about the Nobel prizes.

33. B. Aubrey Fisher, *Small group decision-making: Communication and the group process* (New York, 1974), p. 274.

34. For a discussion of agreement and consensus as the test of "good" decisions in policy making, see C. Lindblom, "The science of 'muddling through,' " *Public Administration Review* 19 (1956), pp. 79–88.

35. H. G. Söderbaum to O. Widman, August 15, 1905, Widman Collection, UUB.

36. This was not so much the case for the sections than for the Academy as a whole. After the reform of 1904, committee representatives made up half of the membership of the physics and chemistry sections.

37. G. Heckscher, *Svensk statsförvaltning i arbete*, Stockholm (1958), pp. 150–3.

38. Aubrey Fisher, *Small group decision-making*, pp. 140–5.

39. Garfinkel's studies of the procedures of jurors' decisions provide interesting parallels with those of the Nobel committees. In comparing decisions in the jury room with those made in everyday life, he suggests that the former differ mainly in that they bear on "the work of assembling the 'corpus' which serves as grounds for inferring the correctness of the verdict." It was only when the outcome was settled that the jurors "went back to find the 'why,' the things that led up to the outcome, and then in order to give their decisions some order which, namely, is the 'officialness' of the decision." (H. Garfinkel, *Studies in ethnomethodology* [Englewood Cliffs, N.J., 1967], pp. 110–15.)

40. 1903, 1906, 1909, 1910, 1912, 1913.

41. *Utkast, KU, kemi*, August 31, 1912.

42. Ångström served from 1904 until his death in 1910, when Granqvist took over. Hasselberg was the first chairman of the physics committee. It is significant that he should have had to give up the chairmanship following the controversy over Arrhenius's call to Berlin in 1904, probably because Hasselberg's actions impaired his ability to lead the committee.

43. H. H. Hildebrandsson, "Knut Ångström," *KVÅb 1913*, pp. 322–3.

44. *Nobelprotokoll, KVA, 4de klassen*, November 30, 1912; *Nobelprotokoll, KVA*, December 4, 1912.

45. Holton, *Thematic origins of scientific thought: Kepler to Einstein*, pp. 261–352.

46. B. Hasselberg to G. E. Hale, July 5, 1907, and December 29, 1907, G. E. Hale Papers, CHP/AIP.

47. *KU, fysik*, 1907; *Les prix Nobel en 1907* (Stockholm, 1909), pp. 13–18. See also Hasselberg, "Om det metriska mått och viktsystemets uppkomst och utveckling," pp. 199–219; D. Livingston Michelson, *The master of light: A biography of A. A. Michelson* (New York, 1973).

48. B. Hasselberg to G. E. Hale, July 5, 1907, G. E. Hale Papers, CHP/AIP.

49. Answering an inquiry from the American astronomer S. Newcomb, in 1900 concerning the inclusion of astrophysics among the fields that could be honored by the physics prize, Hasselberg stated that in his opinion the prize could be awarded for great achievements not only in "pure Physics properly so-called" but also in "the sciences most closely connected with Physics and for the cultivation of which physical methods are employed[;] as [such] Astrophysics and physical chemistry are also to be taken into account" (B. Hasselberg to S. Newcomb, October 26, 1900, Newcomb Collection, LC). Hasselberg's reply was endorsed by the committee (*Protokoll, NK, fysik*, January 8, 1901).

50. H. Wright, *Explorer of the universe: A biography of G. E. Hale* (New York, 1966); R. Michard, "Deslandres, Henri," in *DSB*, Vol. IV, pp. 68–70.

51. *KU, fysik*, 1909.

52. *KU, fysik*, 1913.

53. E. Crawford and R. M. Friedman, "The prizes in physics and chemistry in the context of Swedish science: A working paper," in Bernhard et al., *Science, technology, and society*, pp. 324–7; Friedman, "Nobel physics prize in perspective," pp. 795–8.

54. See the section "Arrhenius's support of atomic theory" in Chapter 5.

55. P. Forman, "The discovery of the diffraction of X-rays by crystals: A critique of the myths," *Archive for History of Exact Sciences*, 6 (1969), pp. 38–71. Forman states: "Although it is often assumed and sometimes asserted, that the theory of the diffraction of X-rays by a molecular space lattice had been worked out in advance of the experimental demonstration, it is almost certain that this was *not* the case" (p. 59).

56. *KU, fysik*, 1914.

57. The importance of Boltzmann's constant *k* had been highlighted in Arrhenius's special report of 1908 (*KU, fysik*, 1908). Granqvist, in *his* special report of 1914, stressed "that both Planck's formula and the constant *h* appearing in it are very important in the study of many different parts of physics" (*KU, fysik*, 1914).

58. Holton applies the term *general theory* primarily to "such few historic and general schemes of thought as the theories of planetary motion, of universal gravitation, of the nuclear atom and the like" (G. Holton and D. Roller, *Foundations of modern physical science* [Reading, Mass., 1958], p. 129).

59. Nagel, "The discussion concerning the Nobel prize for Max Planck," in Bernhard et al., *Science, technology, and society*, pp. 352–76.

60. *Nobelprotokoll, KVA, 4de klassen*, November 28, 1903.

61. *Nobelprotokoll, KVA, 4de klassen*, November 29, 1902.

62. *Nobelprotokoll, KVA*, December 5, 1903. Arrhenius first nominated Bjerknes in 1902 when H. H. Hildebrandsson was coming up for reelection. Unsuccessful, he repeated his proposal the following year and again in 1904 when only ten votes were cast for Bjerknes as against twenty-seven for Hasselberg, who was seeking reelection. (*Nobelprotokoll, KVA*, March 1, 1905.)

63. In the opinion of G. Retzius, Bjerknes's nationality represented a serious obstacle to his election (G. Retzius to S. Arrhenius, December 25, 1904, Arrhenius Collection, SUB-KVAB).

64. That Bjerknes was at the time perhaps the only physicist in Sweden who possessed expertise in mathematical and theoretical physics is illustrated by his remarks in a letter to H. A. Lorentz: "There is nobody in all of *Sweden* with whom I can discuss questions of this kind [Lorentz's electron theory] and I rarely have occasion to travel outside Sweden" (V. Bjerknes to H. A. Lorentz, September 4, 1904, Lorentz Papers, Algemeen Rijksarchief, the Hague).

65. K. Ångström to S. Arrhenius, February 4, 1904, Arrhenius Collection, SUB-

KVAB. In a more disparaging vein, Arrhenius referred to Granqvist as an "electrotechnician" (S. Arrhenius to K. Ångström, February 5, 1904, Ångström Papers, SUB-KVAB).

66. N. Ekholm to S. Arrhenius, March 10, 1910, Arrhenius Collection, SUB-KVAB. In Mittag-Leffler's opinion, Fredholm lost "because the Academy is scared of mathematicians" (G. Mittag-Leffler to I. Fredholm, April 15, 1910, Mittag-Leffler Collection, M.-L. Inst.). As discussed in Chapter 5, Mittag-Leffler felt that the experimentalists on the committee were particularly fearful of higher mathematics, and he therefore sought to dissimulate the role of mathematicians in the campaigns for Poincaré. For a discussion of the attitudes of experimental physicists in general toward mathematics see Pyenson, "Relativity in late Wilhelmian Germany," pp. 140–1.

67. C. W. Oseen, who was to play the most important role in changing committee priorities to theoretical physics after the First World War, adopted this latter attitude. "Given the conditions in Sweden," he wrote to Arrhenius in 1912, "I am inclined to set the boundary as far into physics as possible since by not doing so a void would be created between physics (i.e. experimental physics) and mathematical physics which would harm these two neighboring sciences" (C. W. Oseen to S. Arrhenius, November 13, 1912, Arrhenius Collection, SUB-KVAB). Oseen put this strategy to work in his article "Mechanics and mathematical physics," in J. Guinchard (ed.), *Sweden: Historical and statistical handbook* (Stockholm, 1914), Vol. I, pp. 575–7. To take another example, when Carlheim-Gyllensköld nominated Fredholm (after he himself had been elected), he introduced two kinds of mathematical and theoretical physics: "lower theoretical physics" using "not too difficult mathematics," which, he felt, was already represented on the committee, and "higher mathematical physics" involving "the most stringent methods of modern mathematics," which had no representation. Carlheim-Gyllensköld's statement that "pure experimental physics is more or less represented in all of the members" should also be noted. (*Nobelprotokoll, KVA, 3dje klassen*, November 26, 1910.)

68. *Protokoll, NK, fysik*, September 2, 1911.

69. Arrhenius used his special report on van der Waals to indicate why, in his opinion, the committee could *not* recommend Poincaré for the award. As stated in the letter accompanying the report, this was the passage where he holds up van der Waal's equation concerning the corresponding states of gases and liquids (1873) as a theoretic formulation that had been experimentally verified and, moreover, had borne fruit in new discoveries. It was for this reason that the committee had decided to reward the earlier part of van der Waals's *oeuvre* rather than his more recent works concerning the theory of solutions, which "have not yet proven to be of importance for experimental research." The latter also applied, he felt, to Poincaré's theories in the area of cosmical physics. (S. Arrhenius to G. Granqvist, September 8, 1910, Oseen Collection, SUB-KVAB; *Utkast, KU, fysik*, September 26, 1910.)

70. *KU, fysik*, 1912.

71. *KU, fysik*, 1914.

72. O. Pettersson to N. Wille, October 3, 1905, Wille Collection, OUB.

73. *Protokoll, NK, kemi*, June 4, 1907, and September 7, 1907.

74. *Utkast, KU, kemi*, May 4, 1910, and August 30, 1910.

75. L. F. Haber, *The chemical industry, 1900–1930: International growth and technological change* (Oxford, 1971), pp. 84–90.

76. *KU, kemi*, 1910 (motivation for vote by Klason). In arguing for the importance of nitrogen fixation, Klason drew heavily on the speech that Crookes had made to the British Association for the Advancement of Science and that contained the following oft-quoted phrases: "England and all civilized nations stand in deadly peril of not having enough to eat.... There is a glimmer of light amid this darkness and despondency – the fixation of atmospheric nitrogen." This would be "one of the great discoveries awaiting the ingenuity of chemists." (Quoted in Haber, *The chemical industry, 1900–1930*, p. 84.)

77. See, for example, the special reports by Söderbaum (*KU, kemi*, 1907) and Widman (*KU, kemi*, 1910); see also *KU, kemi*, 1909.
78. *Protokoll, NK, kemi*, August 30, 1910.
79. *Protokoll, NK, kemi*, September 5, 1910.
80. *Nobelprotokoll, KVA, 4de klassen*, October 29, 1910 (memorandum by Å. G. Ekstrand).
81. *Nobelprotokoll, KVA, 4de klassen*, October 29, 1910.
82. *KU, kemi*, 1910.
83. *Nobelprotokoll, KVA, 4de klassen*, October 29, 1910 (memorandum by Ekstrand).
84. *Nobelprotokoll, KVA, 4de klassen*, October 29, 1910 (dissenting opinion by Pettersson).
85. The special report on Frank and Caro commissioned from Klason and Söderbaum was never delivered; instead, Klason presented the committee with the summary of a speech Caro had given at a recent conference in London (*Utkast, KU, kemi*, May 4, 1910). A special report on Schönherr was presented by Widman (*Utkast, KU, kemi*, May 4, 1910). Arrhenius, who had been asked to carry out the experimental verification of Nernst's heat theorem, had declared that given the lack of resources at the Nobel Institute for Physical Chemistry, this could not be done until the fall (in fact, it never materialized). His report on Nernst consisted instead of a summary of recent research showing discrepancies between the values for specific heats predicted by Nernst and those obtained from experiments with different substances (*KU, kemi*, 1910). No special report was commissioned on Wallach either in 1910 or previously.
86. As discussed in Chapter 5, Arrhenius's support of Werner probably helped overcome the hesitancies that the committee previously had expressed with respect to Werner's theory of the bounding of atoms in complex compounds (*KU, kemi*, 1911, 1912). The case of T. W. Richards was different. In fact, here the committee's turnabout was so manifest that Widman felt obliged to provide an explanation in the general report of 1914. In the past, the report stated, he (and others) had doubted that works of pure precision such as Richards's measurements of atomic weights should be rewarded with the prize. He still had the same doubts since it was not yet clear that the results of Richards's work would be of general interest. If he now supported Richards, it was chiefly because of the widespread opinion in the "world of physical chemistry" that Richards deserved the prize on account of his work on atomic weights.
87. *KU, kemi*, 1912.

CHAPTER 7 *The prizes, the public, and the scientific community*

1. These figures are based on the clippings that the Nobel Foundation received from the French agency that monitored the world press. Given the location of the agency, it may have covered the French press more extensively than that of other countries.
2. Frédéric Passy, the founder of the first French peace society, shared the peace prize of 1901 with Henri Dunant, a Swiss, who had founded the International Committee of the Red Cross.
3. *L'Ouest* (November 28, 1911).
4. Rutherford, who had been unofficially told about his Nobel prize by Arrhenius, complained bitterly about speculation in the British press. "It is hopeless contradicting the report in each paper," he wrote, adding in a postscript: "I would like to put the offending reporter in hell with a slow fire under him" (E. Rutherford to S. Arrhenius, November 24, 1908, Arrhenius Collection, SUB-KVAB.)
5. *Le Petit Journal* (January 22, 1902).
6. See, for example, *Die Welt am Montag* (December 11, 1905); *Rheinisch-Westphälische Zeitung* (December 11, 1905).
7. *Vossische Zeitung* (August 20, 1911).

8. *Dagens Nyheter* (October 28, 1914). See also W. Ostwald, "Deutsche Organisation und Wissenschaft," *Die Umschau* (September 25, 1915).
9. *Nobelprotokoll, KVA*, November 27, 1915.
10. Some critical voices were nevertheless raised against the prizes, as, for instance, when the well-known British journalist W. Stead attacked the institution for, among other things, rewarding men who were resting on their laurels long before they became Nobel laureates (*The Independent* [May 9, 1907]). Similar criticisms were later heard in Germany (*Vossische Zeitung* [September 17, 1913]; *Frankfurter Zeitung* [September 18, 1913]).
11. *Evening Transcript* (December 14, 1904).
12. *Cosmopolitan Magazine*, 21 (September 1906).
13. *The Times* (London) (January 6, 7, 8, 1902). Among the authors of letters were E. Gosse, librarian to the House of Lords and a well-known literary figure, and S. P. Thompson, a physicist and active nominator for the science prizes.
14. The announcement of the discovery of X-rays in late 1895 had of course elicited a public response that was probably unsurpassed in the history of modern science. See Nitske, *The life of Wilhelm Conrad Röntgen*, pp. 187 ff.
15. "Marconi is sure to receive the physiological prize [*sic*]," reported the *San Francisco Chronicle* (November 2, 1902), "every scientific society in Europe and the United States having proposed him for this honour." The same view was expressed in *La République* (October 14, 1902), the *Observer* (October 19, 1902), and the *Daily Chronicle* (October 20, 1902), as well as in a number of Italian journals.
16. Reid, *Marie Curie*, pp. 122–33.
17. *Le Figaro* (December 13, 1903); *La Semaine* (December 20, 1903).
18. *La Liberté* (November 15, 1903). The assertion that the Academy of Sciences had passed over the Curies in dispensing such rewards as prizes and grants is not correct. In 1901, Pierre Curie had received one of the most important prizes that the Academy had to offer – the *prix Lacaze* of 10,000 francs. Marie Curie had been the beneficiary of subsidies from grant-giving funds on three different occasions (1898, 1900, 1902) as well as being honored with the *médaille Berthelot* (1902). At the time of their Nobel prize, though, neither of the Curies was a member of the Academy.
19. Even when someone had been considered as sure a prizewinner as Planck in 1908, once the formal announcement had been made, his name vanished from the papers practically overnight and no questions were asked about this candidacy.
20. *Nobelprotokoll, KVA*, October 24, 1903.
21. G. Mittag-Leffler, "Diaries," July 27, 1922, Mittag-Leffler Papers, KB. The statement was made by Gullstrand.
22. L. Badash, *Radioactivity in America: Growth and decay of a science* (Baltimore and London, 1979), pp. 11–12.
23. Reid, *Marie Curie*, pp. 118–24.
24. Hiebert, "Historical remarks on the discovery of argon," pp. 11–19.
25. See, for example, *New York Tribune* (February 21, 1907, and December 13, 1908).
26. The following extract gives the moralizing tone of Arrhenius's letters: "Honor and the esteem which you have for our Academy as well as for science and for your country . . . demand that you refrain from coming here under the present circumstances" (S. Arrhenius to M. Curie, December 1, 1911). Copies of this and other letters exchanged between Arrhenius and Mme. Curie on this subject as well as a letter by P. Langevin explaining his role in the "affair" are found in the Mittag-Leffler Collection, M.-L. Inst. See also Reid, *Marie Curie*, pp. 196–214.
27. *Popular Science Monthly* (January 1907). This article was reprinted in *Science*, 26 (August 16, 1907).
28. *Protokoll, NK, kemi*, March 25, 1904; *Protokoll, gem. sam.*, September 20, 1903.
29. *Svenska Dagbladet* (December 10, 1910). Among the science laureates, responses were received from Lenard, H. A. Lorentz, Ramsey, Rutherford, and J. J. Thomson.

30. A member of the French Academy of Sciences went so far as to convey his condolences to one of the Swedish committee members "on account of the heavy responsibility of awarding a prize of this importance that has fallen on to you" (E. Mascart to H. H. Hildebrandsson, November 19, 1900, Hildebrandsson Collection, UUB).

31. E. Rutherford to S. Arrhenius, November 12, 1908, Arrhenius Collection, SUB-KVAB.

32. B. Schroeder-Gudehus, "Division of labour and the common good: The International Association of Academies, 1899–1914," in Bernhard et al., *Science, technology, and society*, p. 5.

33. Scientists took pride in the prizes awarded to their compatriots because they reflected glory on the scientific enterprise of the country. For example, in 1912, the prize awarded Grignard and Sabatier was used by a prominent French scientist to refute the thesis that chemistry was a predominantly German science (*Le Temps* [November 28, 1912]). For a general discussion of the idea that science had certain national characteristics, see H. Paul, *The sorcerer's apprentice: The French scientist's image of German science, 1840–1919* (Gainesville, Fla., 1972).

34. This mingling of nationalist and internationalist values is brought out by B. Schroeder-Gudehus, *Les scientifiques et la paix: La communauté scientifique internationale au cours des années 20* (Montréal, 1978), pp. 41–62. As discussed in Chapter 2, it was also the basis for the support given the institution of the Nobel prizes by Swedish elites.

35. Schroeder-Gudehus, "Division of labour and the common good," p. 10.

36. Despite the fact that collaborators hardly ever shared in the awards – the most notable exclusion being that of Soddy in 1908 – there was only one formal protest by a scientist who felt that he had been deprived of his just share of the prize. It was made by J.-B. Senderens, Sabatier's closest collaborator and coauthor of his award-winning work. It is significant that Senderens used as one of his arguments the fact that he and Sabatier had shared the *prix Jecker* of the French Academy of Science in 1905. (*Protokoll, NK, kemi*, January 30, 1913 [Appendix B, "Aperçu historique sur le procédé Sabatier-Senderens."])

37. In Forman et al., "Physics *circa* 1900," p. 76, donations of Nobel prizes are considered sufficiently important as a source of financing to figure as a separate category (Table C.4, "Grants and prizes of leading academies *circa* 1900"). To cite some examples of such donations in France and Great Britain: The Curies used part of their prize to hire a laboratory assistant; Sabatier contributed money from his to build the new chemistry institute at the University of Toulouse; Rayleigh's prize helped finance a large new wing at the Cavendish Laboratory in Cambridge.

38. *Svenska Dagbladet* (December 10, 1910).

39. W. Fellchenfeld, "Der Nobelpreis ein Kulturförderer?" *Medizinische Klinik*, 1 (1911), pp. 1–3.

40. E. Curie, *Madame Curie* (Paris, 1938), pp. 191–5.

41. See, for example, *L'Eclair* (May 19, 1905); *L'Echo de Paris* (December 2, 1909); *L'Opinion* (May 27, 1911).

42. M. J. Nye, "The scientific periphery in France: The Faculty of Science at Toulouse (1880–1930)," *Minerva*, 13 (1975), pp. 374–403. See also *Le Temps* (November 28, 1912, and December 26, 1912: interview with C. Bayet, director of higher education in the Ministry of Public Instruction citing efforts made to improve research facilities at provincial universities); *L'Illustration* (November 30, 1912).

43. Zuckerman, *Scientific elite*, pp. 25–35.

44. Dolby, "Debates over the theory of solution," p. 391.

45. R. Abegg to S. Arrhenius, December 23, 1901; January 23, 1902, Arrhenius Collection, SUB-KVAB. See also Körber, *Aus dem Briefwechsel Wilhelm Ostwalds*, Vol. II, pp. 168–9.

46. Ostwald and Nernst held the only two chairs specifically designated for physical

chemistry; they were created in Leipzig and Göttingen, respectively, in 1887 and 1894.

47. Badash, *Radioactivity in America*, pp. 54–5; 261–4.
48. A. S. Eve, *Rutherford* (Cambridge, 1939), p. 429.
49. At the time of Rutherford's award, his physicist colleagues noted their satisfaction at "the unexpected recognition by the students of that science [i.e., chemistry] of the great services which might be rendered to them by researches of purely physical origin" (Cavendish Laboratory, *A history of the Cavendish Laboratory, 1871–1910* [London, 1910], p. 222).
50. See, for example, Heilbron, "Fin-de-siècle physics"; A. Hermann, "Discussion," in Bernhard et al., *Science, technology, and society*, pp. 51–73, 116–18.
51. M. Planck, "New paths of physical knowledge," *Philosophical Magazine*, 28 (1914), p. 61.

Epilogue and conclusions

1. *Nobelprotokoll, KVA*, November 25, 1914.
2. Schroeder-Gudehus, *Les scientifiques et la paix*, pp. 63–97.
3. *Nobelprotokoll, KVA*, November 27, 1915.
4. *Nobelprotokoll, KVA*, November 11, 1915.
5. V. Bjerknes to F. Nansen, January 18, 1902, Nansen Collection, OUB.
6. V. Bjerknes to H. Bjerknes, February 26, 1904, Bjerknes Collection, OUB.
7. Salomon-Bayet, "Bacteriology and Nobel prize selections, 1901–1920," in Bernhard et al., *Science, technology, and society*, pp. 380–6.

Bibliography

The bibliography lists the published works (as well as unpublished doctoral dissertations) cited in the notes. A few publications that are only of incidental interest have been omitted from the bibliography.

The major reference work providing information on the careers and achievements of all the Nobel laureates from this time and practically all the candidates is the *Dictionary of Scientific Biography* (New York: Scribner's, 1970–1978, 15 vols.). Since it would have been too cumbersome to cite the *DSB* with respect to all scientists mentioned in the present volume, the articles included in the bibliography are only those to which references were made in the text.

The bibliography is divided into two parts. The first part lists sources in all languages except Swedish. The second part lists those in the Swedish language. The Swedish publications are listed separately not only because their language renders them inaccessible to many readers but also because such a listing provides to those who know Swedish an overview of the literature pertaining to the Nobel institution that exists only in that language.

NON-SWEDISH SOURCES

Amundsen, L. *Det Norske Videnskaps-Akademi i Oslo, 1857–1957.* 2 vols. Oslo: Aschehoug, 1957–1960.
Anderson, D. *The discovery of the electron: The development of the atomic concept of electricity.* Princeton: Van Nostrand, 1964.
Ångström, K. "Contributions à la connaissance du dégagement de chaleur du radium," *Arkiv för Matematik, Astronomi, och Fysik*, 1–2 (1904–1905), 1–7, 523–8.
Arrhenius, G. "Svante Arrhenius' contribution to earth science and cosmology," In *Svante Arrhenius till 100-årsminnet av hans födelse*, pp. 101–15. (K. Vetenskapsakademiens årsbok, 1959 Bilaga). Stockholm: Almqvist & Wiksell, 1959.
Arrhenius, S. "Electrolytic dissociation versus hydration," *Philosophical Magazine*, 28 (1889), 30–1.
   Theorien der Chemie. 2nd, rev. ed. Leipzig: Akademie-Verlagsgesellschaft, 1909.

Theories of solutions. Silliman Memorial Lectures. New Haven: Yale University Press, 1912.

Worlds in the making. London: Harper and Bros., 1908.

Aubrey Fisher, B. *Small group decision-making: Communication and the group process.* New York: McGraw-Hill, 1974.

Badash, L. *Radioactivity in America: Growth and decay of a science.* Baltimore and London: Johns Hopkins University Press, 1979.

"Rutherford, Ernest." In *Dictionary of scientific biography*, Vol. XII, pp. 26–36. New York: Scribner's, 1975.

Berman, M. *Social change and scientific organization: The Royal Institution, 1799–1844.* London: Heinemann, 1978.

Bernhard, C. G., Crawford, E., and Sörbom, P., eds. *Science, technology, and society in the time of Alfred Nobel.* Oxford: Pergamon Press, 1982.

Bundy, F. P., Hall, H. T., Strong, H. M., and Wentorf, R. H. "Man-made diamonds," *Nature*, 176 (1955), 51–5.

Cavendish Laboratory (University of Cambridge). *A history of the Cavendish Laboratory, 1871–1910.* London: Longmans, 1910.

Cole, J. R., and Cole, S. *Social stratification in science.* Chicago: University of Chicago Press, 1973.

Comberousse, C. de. *J. B. Dumas (1800–1884).* Paris: Publications du Journal Le Génie Civil, 1884.

Combs, H. *Kill Devil Hill: Discovering the secrets of the Wright Brothers.* Boston: Houghton Mifflin, 1979.

Crawford, E. "Arrhenius, the atomic hypothesis, and the 1908 Nobel prizes in physics and chemistry," *Isis*, 75 (1984), forthcoming.

"The prize system of the Academy of Sciences, 1850–1914." In R. Fox and G. Weisz, eds., *The organization of science and technology in France, 1808–1914*, pp. 283–307. Paris and London: Maison des Sciences de l'Homme and Cambridge University Press, 1980.

Crawford, E., and Friedman, R. "The prizes in physics and chemistry in the context of Swedish science: A working paper." In C. G. Bernhard, E. Crawford, and P. Sörbom, eds., *Science, technology, and society in the time of Alfred Nobel*, pp. 311–31. Oxford: Pergamon Press, 1982.

Crawford, E., and MacLeod, R. *The Nobel prizes in physics, 1901–1916: A report on archival materials.* Paris and Brighton: Ecole des Hautes Etudes en Sciences Sociales and University of Sussex, 1976.

Crosland, M. "From prizes to grants in the support of scientific research in France in the nineteenth century: The Montyon legacy," *Minerva*, 17 (1979), 355–80.

Curie, E. *Madame Curie.* Paris: Gallimard, 1938.

Darboux, G. *Eloges académiques et discours.* Paris: Hermann, 1912.

D'Estournelles de Constant, J., Painleve, P., and Bouttieaux, V.-P. *Pour l'aviation.* Paris: Librairie Aéronautique, 1909.

Dictionary of scientific biography. C. Gillispie, ed. 15 vols. New York: Scribner's, 1970–1978.

Dolby, R. G. A. "Debates over the theory of solution: A study of dissent in physical chemistry in the English-speaking world in the late nineteenth and early twentieth centuries," *Historical Studies in the Physical Sciences*, 7 (1976), 297–404.

Euler, H. von. "Svante Arrhenius, 1859–1927." In S. Lindroth, ed., *Swedish men of science, 1650–1950*, pp. 226–38. Stockholm: Almqvist & Wiksell, 1952.

Eve, A. S. *Rutherford: Being the life and letters of the Rt. Hon. Lord Rutherford, O.M.* Cambridge: Cambridge University Press, 1939.

Feather, N. "Some episodes of the alpha-particle story, 1903–1977." In M. Bunge and W. R. Shea, eds., *Rutherford and physics at the turn of the century*, pp. 74–88. New York: Dawson and Science History Publications, 1979.

Fellchenfeld, W. "Der Nobelpreis ein Kulturförderer?" *Medizinische Klinik*, 1 (1911), 1–3.

Forman, P. "The discovery of the diffraction of X-rays by crystals: A critique of the myths," *Archive for History of Exact Sciences*, 6 (1969), 38–71.

Forman, P., Heilbron, J. L., and Weart, S. "Physics *circa* 1900: Personnel, funding, and productivity of the academic establishments," *Historical Studies in the Physical Sciences*, 5 (1975), whole issue.

Fox, R. "The rise and fall of Laplacian physics," *Historical Studies in the Physical Sciences*, 4 (1974), 89–136.

Fredga, A. "Chemistry at Uppsala until the beginning of the twentieth century." In *Faculty of Science at Uppsala University – Chemistry*. (Acta universitatis upsaliensis, Uppsala University 500 years, No. 9), pp. 3–12. Uppsala: Almqvist & Wiksell, 1976.

Friedman, R. M. "Constituting the polar front, 1919–1920," *Isis*, 73 (1982), 343–62.

"Nobel physics prize in perspective," *Nature*, 292 (1981), 793–8.

Friedrich, W., Knipping, P., and von Laue, M. "Interferenz-Erscheinungen bei Röntgenstrahlen," *Bayerische Akademie der Wissenschaften zu München, Sitzungsberichte matematische-physische Klasse*, 42 (1912), 303–22.

Gab [pseud.]. *Monsieur Osiris*. Paris: Eugène Figuière, 1911.

Garfinkel, H. *Studies in ethnomethodology*. Englewood Cliffs, N.J.: Prentice-Hall, 1967.

Gauja, P. *Les fondations de l'Académie des sciences (1881–1915)*. Hendaye, France: Imprimerie de l'Observatoire d'Abbadia, 1917.

Gizycki, R. von. "Centre and periphery in the international scientific community: Germany, France, and Great Britain in the nineteenth century," *Minerva*, 11 (1973), 474–94.

Haber, L. F. *The chemical industry, 1900–1930: International growth and technological change*. Oxford: Clarendon Press, 1971.

Hahn, R. *The anatomy of a scientific institution: The Paris Academy of Sciences, 1661–1803*. Berkeley: University of California Press, 1971.

Harnack, A. *Geschichte der Königlich Preussischen Akademie der Wissenschaften zu Berlin*. 4 vols. Berlin: Reichsdruckerei, 1900.

Hasselberg, B. "Untersuchungen über die Spectra der Metalle im elektrischen Flambogen: 1–8," *K. Vetenskapsakademiens Handlingar*, 26–45 (1894–1910).

Heilbron, J. *"Fin-de-siècle* physics." In C. G. Bernhard, E. Crawford, and P. Sörbom, eds., *Science, technology, and society in the time of Alfred Nobel*, pp. 51–73. Oxford: Pergamon Press, 1982.

H. G. J. Moseley: The life and letters of an English physicist, 1887–1917. Berkeley: University of California Press, 1974.

Hellsten, U. "Ivar Fredholm, 1866–1927." In S. Lindroth, ed., *Swedish men of science, 1650–1950*, pp. 256–63. Stockholm: Almqvist & Wiksell, 1952.

Hermann, A. "Discussion." In C. G. Bernhard, E. Crawford, and P. Sörbom, eds., *Science, technology, and society in the time of Alfred Nobel*, pp. 116–18. Oxford: Pergamon Press, 1982.

Hiebert, E. "Historical remarks on the discovery of argon, the first noble gas." In H. H. Hyman, ed., *Noble-gas compounds*, pp. 3–20. Chicago: Chicago University Press, 1963.

"Nernst and electrochemistry." In G. Dubpernell and J. H. Westbrook, eds., *Selected topics in the history of electrochemistry*, pp. 180–200. Princeton: Electrochemical Society, 1978.

"Nernst, Herman Walther." In *Dictionary of scientific biography*, Vol. XV, Suppl. 1, pp. 432–53. New York: Scribner's, 1978.

"The state of physics at the turn of the century." In M. Bunge and W. R. Shea, eds., *Rutherford and physics at the turn of the century*, pp. 3–22. New York: Dawson and Science History Publications, 1979.

Hiebert, E., and Körber, H.-G. "Ostwald, Friedrich Wilhelm." In *Dictionary of scientific biography*, Vol. XV, Suppl. 1, pp. 455–69. New York: Scribner's, 1978.

Hirosige, T. "Electrodynamics before the theory of relativity, 1890–1905," *Japanese Studies in the History of Science*, 5 (1966), 1–49.

Historical studies in the physical sciences, 3–11 (1971–1980).

Holt, N. "A note on Wilhelm Ostwald's energism," *Isis*, 61 (1970), 386–9.

Holton, G. *Thematic origins of scientific thought: Kepler to Einstein*. Cambridge: Harvard University Press, 1973.

Holton, G., and Roller, D. *Foundations of modern physical science*. Reading, Mass.: Addison-Wesley, 1958.

Ihde, A. *The development of modern chemistry.* New York: Harper & Row, 1964.
Kirchhoff, G. *Vorlesungen über matematische Physik, Vol. I: Mechanik.* 2nd ed. Leipzig: Teuber, 1877.
Kirsten, C., and Körber, H. G., eds. *Physiker über Physiker: Wahlvorschläge zur Aufnahme von Physikern in die Berliner Akademie, 1870 bis 1929.* Berlin: Akademie-Verlag, 1975.
Knight, D. M. *Atoms and elements: A study of theories of matter in England in the nineteenth century.* London: Hutchinson, 1967.
Körber, H.-G., ed. *Aus dem wissenschaftlichen Briefwechsel Wilhelm Ostwalds, Vol. II: Briefwechsel mit Svante Arrhenius und Jacobus Hendricus van't Hoff.* Berlin: Akademie-Verlag, 1969.
Kuhn, T. S. *Black-body theory and the quantum discontinuity, 1894–1912.* New York: Oxford University Press, 1978.
Liljestrand, G. "The prize in physiology or medicine." In H. Schück et al. *Nobel: The man and his prizes,* pp. 131–343. 2nd ed. Amsterdam: Elsevier, 1962.
Lindblom, C. "The science of 'muddling through,' " *Public Administration Review,* 19 (1956), 79–88.
Lindroth, S. *A history of Uppsala University, 1477–1977.* Stockholm: Almqvist & Wiksell, 1976.
Lindroth, S., ed. *Swedish men of science, 1650–1950.* Stockholm: Almqvist & Wiksell, 1952.
Livingston Michelson, D. *The master of light: A biography of A. A. Michelson.* New York: Scribner's, 1973.
McCormmach, R. "Editor's foreword," *Historical Studies in the Physical Sciences,* 3 (1971), ix–xxiv.
"Lorentz, Hendrick Antoon." In *Dictionary of scientific biography,* Vol. VIII, pp. 487–500. New York: Scribner's, 1973.
McGucken, W. *Nineteenth century spectroscopy: Development of the understanding of spectra, 1802–1897.* Baltimore and London: Johns Hopkins University Press, 1969.
MacLeod, R. "Of medals and men: A reward system in Victorian science, 1826–1914," *Notes and Records of the Royal Society of London,* 26 (1971), 81–105.
"The support of Victorian science: The endowment of the research movement in Great Britain, 1868–1900," *Minerva,* 9 (1971), 197–230.
Maier, C. L. "The role of spectroscopy in the acceptance of an internally structured atom, 1860–1920." Ph.D. dissertation, University of Wisconsin, 1964.
Maindron, E. *Les fondations de prix à l'Académie des Sciences: Les lauréats de l'Académie, 1714–1880.* Paris: Gauthier-Villars, 1881.
Malley, M. "The discovery of atomic transmutation: Scientific styles and philosophies in France and Britain," *Isis,* 70 (1979), 213–23.
Melsen, A. G. van. *From atmos to atom.* Pittsburgh: Duquesne University Press, 1952.
Mendelssohn, K. *The world of Walther Nernst: The rise and fall of German science.* London: Macmillan, 1973.
Merton, R. K. *The sociology of science: Theoretical and empirical investigations.* Chicago: University of Chicago Press, 1973.
Merz, J. T. *A history of European thought in the nineteenth century,* Part I: *Scientific thought,* 2 vols. Gloucester, Mass.: Peter Smith, 1976. (Repr. of orig. ed., 1904–1912.)
Michard, R. "Deslandres, Henri." In *Dictionary of scientific biography,* Vol. IV, pp. 68–70. New York: Scribner's, 1971.
Morrel, J., and Thackray, A. *Gentlemen of science: Early years of the British Association for the Advancement of Science.* Oxford: Clarendon Press, 1981.
Moseley, H. G. J. "The high frequency spectra of the elements: Parts I and II," *Philosophical Magazine,* 26 (1913), 1024–34; 27 (1914), 703–13.
Nagel, B. "The discussion concerning the Nobel prize for Max Planck." In C. G. Bernhard, E. Crawford, and P. Sörbom, eds., *Science, technology, and society in the time of Alfred Nobel,* pp. 352–76. Oxford: Pergamon Press, 1982.
Nitske, W. R. *The life of Wilhelm Conrad Röntgen, discoverer of the X-ray.* Tucson: University of Arizona Press, 1971.
Nobel Foundation. *Directory, 1981–1982.* Stockholm: Nobel Foundation, 1981.
Nobel lectures, including presentation speeches and laureates' biographies: Chemistry, 1901–1921. Amsterdam: Elsevier, 1967.

Nobel lectures, including presentation speeches and laureates' biographies: Physics, 1901–1921. Amsterdam: Elsevier, 1967.

Nye, M. J. "Gustave Le Bon's black light: A study in physics and philosophy in France at the turn of the century," *Historical Studies in the Physical Sciences*, 4 (1974), 163–95.

Molecular reality: A perspective on the scientific work of Jean Perrin. London and New York: Macdonald and American Elsevier, 1972.

"N-rays: An episode in the history and psychology of science," *Historical Studies in the Physical Sciences*, 11 (1980), 125–56.

"The scientific periphery in France: The Faculty of Science at Toulouse (1880–1930)," *Minerva*, 13 (1975), 374–403.

Oseen, C. W. "Mechanics and mathematical physics." In J. Guinchard, ed., *Sweden: Historical and statistical handbook*, Vol. I, pp. 612–14. 2nd ed. Stockholm: Norstedt, 1914.

Ostwald, W. Review of *Studien zur Lehre von den kolloiden Lösungen* by T. Svedberg, *Zeitschrift für physikalische Chemie*, 64 (1908), 508–9.

Partington, J. R. *A history of chemistry*, Vol. IV. London: Macmillan, 1972.

Paul, H. *The sorcerer's apprentice: The French scientist's image of German science, 1840–1919.* Gainesville: University of Florida Press, 1972.

Pihl, M. "Bjerknes, Vilhelm Frimann Koren." In *Dictionary of scientific biography*, Vol. II, pp. 167–9. New York: Scribner's, 1970.

Planck, M. "New paths of physical knowledge," *Philosophical Magazine*, 28 (1914), 60–71.

Poincaré, H. "Analyse des travaux scientifiques de Henri Poincaré faite par lui-même," *Acta Mathematica*, 38 (1921), 3–135.

Electricité et optique: La lumière et les théories électrodynamiques. Leçons professées à la Sorbonne en 1888, 1890, et 1899. 2nd. rev. ed. Paris: Carré et Naud, 1901.

"Relations entre la physique expérimentale et la physique mathématique." In C. E. Guillaume and L. Poincaré, eds., *Rapports présentés au Congrès International de Physique, Paris, 1900*, Vol. I, pp. 1–29. Paris: Gauthier–Villars, 1900–1901.

Post, H. R. "Atomism, 1900: I and II," *Physics Education*, 3 (1968), 225–32, 307–12.

Les Prix Nobel, 1901–1918. 14 vols. Stockholm: Norstedt, 1904–1920.

Pyenson, L. "Relativity in late Wilhelmian Germany: The appeal to a preestablished harmony between mathematics and physics," *Archive for History of Exact Sciences*, 27 (1982), 137–55.

Ramsay, W. "The electron as an element," *Journal of the Chemical Society: Transactions*, 93 (1908), 774–88.

Ramser, H. "Warburg, Emil Gabriel." In *Dictionary of scientific biography*, Vol. XIV, pp. 170–2. New York: Scribner's, 1976.

Rayleigh, J. W. S., and Ramsay, W. "Argon, a new constituent of the atmosphere," *Philosophical Transactions of the Royal Society* (A), 186 (1896), 187–241.

The Record of the Royal Society of London. London: Harrison, 1897.

Reid, R. *Marie Curie.* London: Collins, 1974.

Riesenfeld, E. *Svante Arrhenius.* Leipzig: Akademische Verlagsgesellschaft, 1931.

Root-Bernstein, R. "The Ionists: Founding physical chemistry, 1872–1890." Ph.D. dissertation, Department of History, Program in History and Philosophy of Science, Princeton University, 1980.

Salomon-Bayet, C. "Bacteriology and Nobel prize selections, 1901–1920." In C. G. Bernhard, E. Crawford, and P. Sörbom, eds., *Science, technology, and society in the time of Alfred Nobel*, pp. 377–400. Oxford: Pergamon Press, 1982.

Schonland, B. *The atomists (1805–1933).* Oxford: Clarendon Press, 1968.

Schroeder-Gudehus, B. "Division of labour and the common good: The International Association of Academies, 1899–1914." In C. G. Bernhard, E. Crawford, and P. Sörbom, eds., *Science, technology, and society in the time of Alfred Nobel*, pp. 3–20. Oxford: Pergamon Press, 1982.

Les scientifiques et la paix: La communauté scientifique internationale au cours des années 20. Montreal: Les Presses de l'Université de Montréal, 1978.

Schück, H., et al. *Nobel: The man and his prizes.* 2nd ed. Amsterdam: Elsevier, 1962.

Siegbahn, M. "Janne Rydberg, 1854–1919." In S. Lindroth, ed., *Swedish men of science, 1650–1950*, pp. 214–18. Stockholm: Almqvist & Wiksell, 1952.

Simon, H. *Administrative behaviour.* 3rd ed. London and New York: Free Press, 1976.
Sitzungsberichte der Königlich Preussischen Akademie der Wissenschaften, 1914. Berlin: Verlag der Königlichen Akademie der Wissenschaften, 1914.
Sohlman, R. "Alfred Nobel and the Nobel Foundation." In H. Schück *et al. Nobel: The man and his prizes*, pp. 15–72. 2nd ed. Amsterdam: Elsevier, 1962.
The legacy of Alfred Nobel: The story behind the Nobel prizes. London: The Bodley Head, 1983.
Sundelöf, L.-O. and Pedersen, K. O. "Physical chemistry." In *Faculty of Science at Uppsala University – Chemistry.* (Acta universitatis upsaliensis, Uppsala University 500 years, No. 9), pp. 153–98. Uppsala: Almqvist & Wiksell, 1976.
Trenn, T. *The self-splitting atom: The history of the Rutherford–Soddy collaboration.* London: Taylor & Francis, 1977.
Van Den Handel, J. "Kammerlingh Onnes, Heike." In *Dictionary of scientific biography*, Vol. VII, pp. 220–2. New York: Scribner's, 1973.
Westgren, A. "The chemistry prize." In H. Schück, *et al. Nobel: The man and his prizes*, pp. 345–438. 2nd ed. Amsterdam: Elsevier, 1962.
Whittaker, E. *A history of theories of aether and electricity.* 2 vols. Rev. ed. London: Nelson, 1953.
Wright, H. *Explorer of the universe: A biography of G. E. Hale.* New York: Dutton, 1966.
Yearbook of the Royal Society of London, 1907. London: Harrison, 1907.
Ziman, J. *Reliable knowledge.* Cambridge: Cambridge University Press, 1978.
Zuckerman, H. *Scientific elite: Nobel laureates in the United States.* New York and London: Free Press, 1977.

SWEDISH SOURCES

Ångstrom, K. "Fysiska institutionen." In *Uppsala Universitet, 1872–1897: Festskrift*, Vol. II, pp. 148–55. Uppsala: Akademiska Boktryckeriet, 1897.
Arrhenius, O. "Svante Arrhenius: Det första kvartsseklet." In *Svante Arrhenius till 100-årsminnet av hans födelse*, pp. 43–64. (K. Vetenskapsakademiens årsbok 1959. Bilaga). Stockholm: Almqvist & Wiksell, 1959.
Arrhenius, S. "Atomlärans framgångar," *Svensk Kemisk Tidskrift*, 20 (1908), 172–8.
"Carl Lindhagen och Stockholms Högskola." In *Carl Lindhagen 60 år*, pp. 41–5. Stockholm: Svenska Andelsförlaget, 1920.
Arrhenius-Wold, A.-L. "Svante Arrhenius och utvecklingsoptimismen." In *Svante Arrhenius till 100-årsminnet av hans födelse*, pp. 65–100. (K. Vetenskapsakademiens årsbok 1959. Bilaga). Stockholm: Almqvist & Wiksell, 1959.
Beckman, A., and Ohlin, P. *Forskning och undervisning i fysik vid Uppsala Universitet under fem århundraden: En kortfattad historik.* (Acta universitatis upsaliensis. Skrifter rörande Uppsala Universitet. C. Organisation och historia, No. 8). Uppsala: Almqvist & Wiksell, 1965.
En befordringsfråga vid Stockholms Högskola. Stockholm: Stockholms Dagblads Tryckeri, 1895.
Carleman, T. "Magnus Gustaf Mittag-Leffler." In *Levnadsteckningar över Kungl. Svenska Vetenskapsakademiens ledamöter*, Vol. VII, pp. 459–71. Stockholm: Almqvist & Wiksell, 1939–1948.
Dahlgren, E. *Kungl. Svenska Vetenskapsakademien: Personförteckningar, 1739–1915.* Stockholm: Almqvist & Wiksell, 1915.
Eriksson, G. *Kartläggarna: Naturvetenskapens tillväxt och tillämpningar i det industriella genombrottets Sverige, 1870–1914.* (Acta Universitatis Umensis, No. 15), Umeå: Nyheternas Tryckeri, 1978.
Euler, H. von. "Sven Otto Pettersson In Memoriam," *Svensk Kemisk Tidskrift*, 53, (1941), 28–32.
Handlingar rörande donationer till K. Vetenskapsakademien eller under dess förvaltning. Stockholm: Norstedt, 1880–1901.
Hasselberg, B. "Om det metriska mått och viktsystemets uppkomst och utveckling,"

K. *Vetenskapsakademiens årsbok 1908*, pp. 199–219. Uppsala and Stockholm: Almqvist & Wiksell, 1908.

"Tobias Robert Thalén," K. *Vetenskapsakademiens årsbok 1906*, pp. 218–40. Uppsala and Stockholm: Almqvist & Wiksell, 1906.

Heckscher, G. *Svensk statsförvaltning i arbete*, 2nd rev. ed. Stockholm: Norstedt, 1958.

Hildebrandsson, H. H. "Knut Ångström," K. *Vetenskapsakademiens årsbok 1913*, pp. 302–27. Uppsala and Stockholm: Almqvist & Wiksell, 1913.

Holmberg, B. "Olof Hammarsten," K. *Vetenskapsakademiens årsbok 1934*, pp. 272–93. Stockholm: Almqvist & Wiksell, 1934.

Karolinska Medikokirurgiska Institutets historia utgiven med anledning av Institutets hundraårsdag. 3 vols. in 2. Stockholm: Isaac Marcus, 1910.

Kungl. Vetenskapsakademiens årsbok (1906–1934). Uppsala and Stockholm: Almqvist & Wiksell, 1906–1934.

Lindhagen, C. *Carl Lindhagens memoarer.* 3 vols. Stockholm: Bonniers, 1936.

Lindroth, S. *Kungl. Svenska Vetenskapsakademiens historia, 1739–1818.* 2 vols. Stockholm: Almqvist & Wiksell, 1967.

Lundgren, A. "Arrhenius om van't Hoff och Erhlich," *Lynchnos* (1975–1976), 85–100.

Molin, K. "Vilhelm Carlheim-Gyllensköld som jordmagnetiker," *Kosmos*, 13 (1935), 5–60.

Ölander, A. "Arrhenius och den elektrolytiska dissociationsteorin." In *Svante Arrhenius till 100-årsminnet av hans födelse*, pp. 5–33. (K. Vetenskapsakademiens årsbok 1959. Bilaga). Stockholm: Almqvist & Wiksell, 1959.

Oseen, C. W. *Atomistiska föreställningar i nutidens fysik: Tid, rum, och materia.* Stockholm: Bonniers, 1919.

Pettersson, O. Review of *Recherches sur la conductibilité galvanique des électrolytes* by S. Arrhenius, *Nordisk revy*, 2 (1884), 204–7. (Repr. in *Svensk Kemisk Tidskrift*, 15 [1903], 208–11.)

"Svante Arrhenius," *Svensk Kemisk Tidskrift*, 15 (1903), 199–208.

Protokoll hållna vid sammanträden för överläggning om Alfred Nobels testamente. Stockholm: Norstedt, 1899.

Ramberg, L. "Oscar Widman," K. *Vetenskapsakademiens årsbok 1931*, pp. 289–334. Stockholm: Almqvist & Wiksell, 1931.

Siegbahn, M. "Gustaf Granqvist," K. *Vetenskapsakademiens årsbok 1924*, pp. 308–13. Stockholm: Almqvist & Wiksell, 1924.

Sohlman, R. "Carl Lindhagen och Nobelstiftelsen." In *Carl Lindhagen 60 år*, pp. 56–60. Stockholm: Svenska Andelsförlaget, 1920.

Ett testamente. Stockholm: Norstedt, 1950.

Statistik över löneförhållandena inom Svenska Träarbetarförbundet för år 1898. Stockholm: Arbetarnes Tryckeri, 1900.

Stockholms Högskola, 1878–1898: Berättelse över Stockholms Högskola utveckling under hennes första tjugoårsperiod på uppdrag av hennes lärarråd utgiven av Högskolans rektor. Stockholm: Norstedt, 1900.

Strindberg, A. *August Strindbergs brev.*, ed. T. Eklund, Vols. VII–XIV. Stockholm: Bonniers, 1961–1974.

Sundin, J. *Främmande studenter vid Uppsala Universitet före andra världskriget.* (Acta universitatis Upsaliensis, Skrifter rörande Uppsala Universitet. C. Organisation och historia, No. 27). Uppsala: Almqvist & Wiksell, 1973.

Torbacke, J. "Affären Staaff–Mittag-Leffler," *Statsvetenskaplig Tidskrift*, 1961, 10–42.

Tunberg, S. *Stockholms Högskolas historia före 1950.* Stockholm: Norstedt, 1957.

Waller, I. "Carl Wilhelm Oseen." In *Levnadsteckningar över Kungl. Svenska Vetenskapsakademiens ledamöter*, Vol. VIII, pp. 121–43. Stockholm: Almqvist & Wiksell, 1949–1954.

Index*

*References to "Prize" denote "Nobel
prize."